Max Delbrück and Cologne

An Early Chapter of German Molecular Biology

Max Delbrück and Cologne

An Early Chapter of German Molecular Biology

Simone Wenkel
University of Cologne, Germany

Ute Deichmann
University of Cologne, Germany & Leo Baeck Institute, UK

NEW JERSEY • LONDON • SINGAPORE • BEIJING • SHANGHAI • HONG KONG • TAIPEI • CHENNAI

Published by

World Scientific Publishing Co. Pte. Ltd.
5 Toh Tuck Link, Singapore 596224
USA office: 27 Warren Street, Suite 401-402, Hackensack, NJ 07601
UK office: 57 Shelton Street, Covent Garden, London WC2H 9HE

Library of Congress Cataloging-in-Publication Data
Max Delbrück and Cologne : an early chapter of German molecular biology / editors, Simone Wenkel, Ute Deichmann.
p. ; cm.
Includes bibliographical references and index.
Text in English and German.
ISBN-13: 978-981-270-547-1 (hardcover : alk. paper)
ISBN-10: 981-270-547-3 (hardcover : alk. paper)
1. Delbrück, Max. 2. Molecular biologists--United States--Biography. 3. Universität zu Köln. Institut für Genetik. 4. Molecular biology--History. I. Wenkel, Simone. II. Deichmann, Ute, 1951–
[DNLM: 1. Delbrück, Max. 2. Universität zu Köln. Institut für Genetik. 3. Academies and Institutes--history. 4. Molecular Biology--history. 5. History, 20th Century. 6. Science--history. QU 24 U58m 2007]
QH31.D434M39 2007
572.8092--dc22

2007016771

British Library Cataloguing-in-Publication Data
A catalogue record for this book is available from the British Library.

Typeset by Stallion Press
Email: enquiries@stallionpress.com

Printed in Singapore.

Preface

When members of the Institute for Genetics at the University of Cologne started to look back on the history of their institute in 2004, they had actually forgotten that its famous founder, Max Delbrück, would have celebrated his 100th anniversary on September 4, 2006. To mark the event, they, that is Maria Leptin, Jonathan Howard, Simone Wenkel and Ute Deichmann, organised a workshop on the early history of the Institute, in April 2005. Most of the founding members participated, and in the end, of course, the timing was ideal. The current volume is an extended proceedings, in which articles based on the talks given at this workshop are supplemented by articles from other scientists and historians, and is in large part a tribute to Max Delbrück's initiative.

Why is the history of the Institute of Genetics worth remembering? To give an appropriate answer, we have to make a historical excursion to the 1930s. Before the Nazis came to power, German scientists were internationally renowned for their novel work in biochemistry, the chemistry of natural products as well as the biochemistry of intermediate metabolism. But there was no pioneering work in biochemistry or molecular biology after the Second World War. Research in bacteria and phage genetics and research into the structure and function of proteins and nucleic acids, later subsumed under the term "molecular biology", started in Germany only around 15 years after it was initiated in the United States and England. By contrast, experimental genetic mutation research had been implemented in Germany in the late 1920s, as soon as Hermann J. Muller published his work on artificial mutagenesis by X-rays; and research on the biochemistry and genetics of tobacco mosaic virus had been started in the late 1930s, immediately after Wendell Stanley published his first promising work on this virus in the United States.

The comparatively late beginning of the "new" molecular biology in Germany can be explained mainly by the following factors, in addition to the general problem of physical destruction and material shortages from which universities were suffering: first, the forced emigration of Jewish scientists after 1933 had led not only to the expulsion of future distinguished molecular biologists, but also to a marked decline of "dynamic biochemistry", a field which contributed greatly to molecular biology in the 1960s. Second, German university structures seriously impeded new and interdisciplinary research. Third, the international isolation and self-isolation of German scientists, a consequence of National Socialism and the Second World War, was an obstacle to the implementation of new fields of research developing elsewhere. It reinforced the conduct of traditional genetic research at German universities, where, until the 1970s, research was conducted almost exclusively on complex organisms.

Max Delbrück, an émigré physicist (he was not Jewish but emigrated for other reasons in 1937) who had founded phage genetics in the late 1930s in the United States, played a decisive role in facilitating the international exchange of mainly young German scientists. Because he maintained strong connections to Germany, in the immediate post-war period he was the only scientist abroad to help the Germans establish modern, mostly molecular, biology and catch up with international developments. It required a new generation of young scientists who had received part of their training in other countries, particularly in the United States, before modern molecular biology became established at German universities and Max Planck Institutes.

Max Delbrück's influence is most visible in the founding of the Institute of Genetics at the University of Cologne — the first molecular genetics institute at a German university. Founded in 1959, the Institute was the first to implement the less hierarchical American organisational structures and research habits in Germany. The Institute quickly gained high international scientific recognition, and in the 1960s was, as Delbrück described it, "on the map of itinerant molecular biologists" from all over the world.

In this volume, 31 contributions by scientists and historians deal with the founding of the Institute; with its founder Max Delbrück; with its unusually liberal atmosphere; and with its internationally oriented research. The majority of the contributions consist of reports and reflections by scientists who experienced the founding phase of the Institute as students or academics. Other essays show perceptions from the outside and compare aspects of the Institute's research and teaching with other institutes — both in Germany and abroad. Contributions by historians of science analyse the historical background and international context of the founding of the Institute and its research. Participants in panel discussions address current questions of science policy, in particular concerning teaching and the impact of the structure of scientific institutions on the quality of research.

Two documents from the 1960s are included in the volume. The first is a reprint of Niels Bohr's lecture "Light and Life" at the inauguration of the Institute, with an introduction by Gunther S. Stent. Delbrück owed his interest in biology to Bohr's 1932 lecture with the same title, in which Bohr expressed the romantic hope that new laws of physics might be found that explain the attributes of life; a hope that did not come true, at least in regard to genetics, as Delbrück later admitted. The second reprint is from Joseph Straub, a botanist in Cologne, who, by convincing Delbrück to come to Cologne and organising the political and financial support, was most instrumental in the founding of the Institute. In his lecture at the Delbrück memorial in 1982 Straub recalls the fascinating time he spent with Max Delbrück in Cologne.

The Editors
Köln, October 2006

Contents

Part I. Introduction

1

A Brief Review of the Early History of Genetics and Its Relationship to Physics and Chemistry

Ute Deichmann

Physicists and formal genetics: from Gregor Mendel to Max Delbrück

Gregor Mendel's *Versuche über Pflanzen-Hybriden* (1866) is recognised as the foundation of what later became the science of genetics. Unlike most other fields of biology, genetics from its beginnings has been close to physics and mathematics. Mendel received an education not only in botany, but also in the physical sciences before he began his experiments on pea hybridisation. His aim to find a "generally applicable law" (governing the formation and development of hybrids) is more reminiscent of the hard sciences than of biology. Unlike contemporary plant breeders Mendel considered it critically important to obtain exact numerical data for each of the traits he looked at. Convinced that there was a numerical order in nature, he succeeded to find equations by which the distribution of pea hybrids' properties in subsequent generations can be adequately described statistically. According to Mendel, discrete and independent factors (later called genes) accounted for the inheritance of each trait of an organism.

His allowance of chance to play a decisive role in natural processes[1] and his assumption that inheritance is "discrete", and not "blending", with discrete factors underlying seemingly continuous phenomena, bear similarities to developments in physics in the late 19th century. Statistical mechanics, the application of statistics to the physical branches of classical mechanics and thermodynamics, was the first physical theory in which probabilistic concepts and the notion of chance played a fundamental role. In this theory, which was founded by Ludwig Boltzmann and, independently, J. Willard Gibbs, the laws of nature only apply to large populations, and predictions are possible not for the causes of isolated events but with a probability for the totality of events. This also held true for Mendel's rules.[2]

For various reasons Mendel's work was largely neglected for 35 years. The "rediscovery" of his rules in 1900 by Carl Correns, Hugo de Vries and Erich Tschermak-Seysenegg marks the beginning of classical, or formal, genetics, which focussed on the transmission of traits from parents to offspring. In 1906 in England William Bateson coined the term "genetics". He developed major concepts and a large part of the early terminology of the emerging science. The term "gene" as the "genotypic" basis of a distinct "phenotypic" trait was introduced by the Danish researcher Wilhelm Johannsen in 1909. His notion of the gene had a far-reaching impact; while realising that the behaviour of genes had something in common with "chemical bodies", he concluded nevertheless that this did not mean that genes themselves were chemical entities. He suggested that the term gene be used merely as an abstraction, "for the time being only

---

[1]His frequent allusion to chance effects can be found, for example, in the following paragraph of his 1866 paper: "This represents the *average* result of the self-fertilization of the hybrids when two differentiating characters are united in them. ... It remains purely a matter of chance which of the two sorts of pollen may fertilize each separate egg cell". Mendel's paper, in its original German and in English translation, is available at www.mendelweb.org.

[2]For a comparison of Mendel's approach with statistical mechanics see Francois Jacob, *The Logic of Life. A History of Heredity* (Princeton, 1993), pp.192–201.

something like a unit of calculation."[3] This gene concept became one of the most powerful abstractions in biology.

In 1910 Thomas Hunt Morgan in the United States observed a white-eyed mutant male among the red-eyed wild-type individuals of his new experimental object, Drosophila. Cross-breeding experiments showed that only males displayed this trait. Morgan concluded that it was sex-linked and its gene was carried on the X-chromosome. Experiments such as this one marked the beginning of a second phase of Mendelian genetics, in which Morgan and his collaborators, Alfred H. Sturtevant, Calvin B. Bridges and Hermann J. Muller, developed the chromosome theory of inheritance, according to which genes are located on chromosomes and transmitted in linkage groups unless crossover occurs. They succeeded to establish gene maps of the four chromosomes.[4] By promoting the belief that the gene is the "basis of life" Muller emphasised the importance of the new science.[5] He considered genes' basic properties to be identical self-replication and heritable mutations, which he regarded as the basis of biological evolution.[6] Already in 1922 he envisaged that geneticists would have to become bacteriologists, biochemists and physicists. But convinced that, at least at the time, "a gene cannot effectively be ground in a mortar or distilled in a retort",[7] he continued to pursue and propagate the indirect methods of formal genetics

---

[3]Wilhelm Johannsen, *Elemente der exakten Erblichkeitslehre* (Jena, 1909), pp. 124–125. The third edition (1926, p. 168) contains the same sentence. The literature on the history of genetics and molecular biology is numerous. Of the recent pulications I mention Raphael Falk, "The gene — a concept in tension," in: Peter Beurton. Raphael Falk, and Hans-Joerg Rheinberger, *The Concept of the Gene in Development and Evolution* (Cambridge, 2000), pp. 317–348. A classic in the history of early molecular biology, covering medical microbiology, virus research, biochemical and biophysical genetics is Robert Olby, *The Path to the Double Helix. The Discovery of DNA* (Seattle, 1974).

[4]Already in 1903 and 1904, respectively, Walter Sutton and Theodor Boveri, looking for a cytological basis of genetics, formulated early versions of the chromosome theory.

[5]Hermann J. Muller, "The gene as the basis of life", *Proceedings of the International Congress of Plant Science*, 1929. Muller gave this speech in 1926.

[6]Hermann J. Muller, "Variation due to change in the individual gene", *American Naturalist* **56** (1922), pp. 32–50.

[7]Ibid.

and, later, genetic radiation studies. His discovery in 1927 of artificial mutagenesis by X-rays opened up a new approach in genetics (see below) and also an important branch of environmental studies.

Morgan and his collaborators endowed genes with a location and some physical existence. But the search for the material nature of the gene was not relevant for this highly successful approach, as the following quotation from Morgan's Nobel lecture, as late as 1934, shows: "There is no consensus of opinion amongst geneticists as to what the genes are — whether they are real or purely fictitious — because at the level at which the genetic experiments lie, it does not make the slightest difference whether the gene is a hypothetical unit, or whether the gene is a material particle". As Max Delbrück later explained, in the mid-1930s genes were still "algebraic units of the combinatorial science of genetics, and it was anything but clear that these units were molecules analysable in terms of structural chemistry".[8]

Classical genetics soon became an exact biological discipline. The physicist Max Delbrück was fascinated with the fact that it was a logically self-contained autonomous exact science which was quantitative without using physical measures such as velocity and mass, in contrast to chemistry which had not obtained independence from physics.[9]

Muller's discovery that X-ray radiation causes mutations led to a collaboration of geneticists and physicists on the investigation of the biological effects of radiation. On Muller's suggestion, Berlin geneticist Nikolai Timoféeff-Ressovsky started collaborating with the physicist Karl G. Zimmer. They were joined by Max Delbrück in the early 1930s. As a theoretical physicist Delbrück was not experienced in experimental radiation research. In the following, I shall present some background information on his becoming a biologist and geneticist.

---

[8]Max Delbrück, "A physicist's renewed look at biology: Twenty years later", *Science* **168** (1970), pp. 1312–1315. According to Delbrück, genes "could have turned out to be submicroscopic steady-state systems, or they could have turned out to be something unanalysable in terms of chemistry, as first suggested by Bohr".

[9]Max Delbrück, "On the nature of gene mutations and gene structure," in Nikolai W. Timoféeff-Ressovsky, Karl G. Zimmer und Max Delbrück, "Über die Natur der Genmutation und der Genstruktur", *Nachrichten von der Gesellschaft der Wissenschaften zu Göttingen (Mathematisch-Physikalische Klasse)* **1** (1935), pp. 189–245.

Delbrück spent three post-doctoral years (1929–32) in England, Switzerland and Denmark. The encounter with a new language and culture in England and the association with Wolfgang Pauli in Switzerland and Niels Bohr in Denmark influenced him deeply. Bohr's speculation that the complementarity argument of quantum mechanics might be applied to the explanation of phenomena of life, summarised in his 1932 lecture "Light and life",[10] aroused Delbrück's interest in biology. Bohr assumed that the laws of biology might be complementary to those of physics and chemistry in a similar way as the "spatial continuity of light propagation" according to electromagnetic theory is complementary to the "atomicity of the light effects", the light quanta. Both features are aspects of one reality, the phenomenon of light, and cannot contradict each other because a closer analysis of each aspect would demand "mutually exclusive experimental arrangements".[11] Bohr regarded the existence of life as an "elementary fact that cannot be explained, but must be taken as a starting point in biology" and concluded with the vitalistic notion that "the asserted impossibility of a physical or chemical explanation of the function peculiar to life" would be "analogous to the insufficiency of the mechanical analysis for the understanding of the stability of atoms".

Bohr's notion of biology stood in stark contrast to the view that Jacques Loeb, a German-American biologist and biochemist, had propagated forcefully from the late 19th century. Loeb asserted that life phenomena can be explained by chemical and physical methods, and that biologists had to use and develop these methods further if they wanted to work scientifically. Already in the 1910s Loeb stimulated the earliest experimental attempts to associate (but not equate) genes and enzymes. As it turned out a few years later, Bohr's hopes, that in biology completely different laws from those in physics might be found, were not fulfilled. Instead, Loeb's "mechanistic" programme culminated in Watson's and Crick's theory of the double helix structure of

---

[10]Bohr's lecture was published in *Nature* **131** (1933), pp. 421–423; 457–459.

[11]Ibid. On the impact of Bahr's complementarity concept see for example Ernst P. Fischer and Carol Lipson, *Thinking about Science. Max Delbrück and the Origins of Molecular Biology*, New York, 1988.

DNA, in which "complementarity" became an entirely chemical concept: the specific pairing of DNA's four bases, through weak forces of hydrogen bonds, according to their chemical structure.

In the 1930s, however, Bohr's romantic views motivated Delbrück and a few other physicists to search for new laws of life. Back in Germany, where he became an assistant to Lise Meitner at the Kaiser Wilhelm Institute (KWI) for Chemistry in 1932, Delbrück founded a discussion group of biologists and physicists. It led to a co-operation with Timoféeff-Ressovsky and Zimmer and the publication of a lengthy and influential joint paper *Über die Natur der Genmutation und der Genstruktur*, which consisted of three separate contributions and a joint conclusion by its authors.[12]

A central point of the paper, often referred to as the "Three-Man-Work", was the physical interpretation of the gene. Delbrück wrote a theoretical section entitled *Atomphysikalisches Modell der Genmutation*, which is acknowledged to be of fundamental importance in the history of molecular biology. Reflecting on the nature of the stability of the gene as a "well-defined association of atoms", Delbrück was the first to develop the notion of genes as something similar to macromolecules. The problem of the two aspects of mutations, change and constancy of the change, was solved by assuming that the stability of the gene was caused by the strength of interatomic forces and its mutation by a quantum jump from one stable configuration to another through certain forms of energy from the outside. Notwithstanding the fact that Delbrück did not expect chemistry to provide the solution for the question of the nature of the gene — for example, he did not relate its stability to a molecular configuration — he presented for the first time a theoretical concept that made genes approachable by physical and chemical means.

His elaboration had strong direct and indirect effects. Impressed with Delbrück's thoughts, Warren Weaver from the Rockefeller Foundation in 1936 offered him a fellowship for a stay in the United States. In 1938 Delbrück's paper motivated the physiologist Salvador Luria, a Jewish refugee from Italy in the United States, to take pains to start collaboration with Max Delbrück. Finally, in 1944, Delbrück's

[12]Timoféeff *et al.*, "Über die Natur".

paper became the central constituent of Erwin Schrödinger's book, *What is Life.*

Unlike other physicists, Delbrück did not continue to deal with radiation experiments in order to understand the nature of genes. As he stated with hindsight, his decision was a right one: the hope "to get at the chemical nature of the gene by means of radiation genetics never materialised. The road to success effectively bypassed radiation genetics".[13] In particular, the application of the "target theory", a theoretical stochastic model of radiation-induced effects, to mutation analysis did not prove fruitful for elucidating properties of the gene, such as its size. Instead, Delbrück was fascinated with the discovery of the American biochemist Wendell Stanley in 1935 that a certain virus, the tobacco mosaic virus (TMV), could be crystallised. Since viruses are able to replicate identically Delbrück concluded that TMV could be used as a primitive model of the gene, and that virus research might be relevant "for a general assessment of the phenomena peculiar to life".[14] Interestingly, already in 1937 he sensed that these phenomena might be simpler than anticipated by Bohr: "One should view replication not as complementary to atomic physics but as a particular trick of organic chemistry".[15]

After 1933, Delbrück was not dismissed from his position at the KWI. However, due to his "misbehaviour" at a Nazi training course for university teachers, he was not immediately accepted for *Habilitation.* Instead of staying in Berlin and trying again to pursue an academic career, he accepted a fellowship provided by the Rockefeller Foundation and in 1937 moved to the United States. After visiting various laboratories he became convinced that neither Drosophila nor TMV were experimental objects suited for studying gene replication. Only when he found a congenial experimental object, phage (bacterial viruses), with which he was able to achieve fast and clear cut quantitative results on phage growth and mutation, did he become a biologist.

Genetic research in phage, as it was initiated by Delbrück, who was soon followed by Salvador Luria and Alfred Hershey, proved successful in tackling problems such as gene mutation and replication. An

[13]Delbrück, "A physicist's renewed view".

[14]Ibid., appendix 1 (Delbrück's notes of 1937).

[15]Ibid.

example is the so-called fluctuation experiment.[16] Luria and Delbrück showed quantitatively that adaptation of bacteria to virus resistance was a result of random mutations occurring prior to contact with the virus and subsequent selection and not, as was widely believed, of directed mutations in a Lamarckian sense. Elie Wollman, a molecular biologist at the Institut Pasteur in Paris, later characterised the novelty of the approach of the phage group as compared with that of earlier researchers on phage:

> "Taking a strictly Cartesian attitude, Delbrück and Luria had swept away the facts and interpretations accumulated by their predecessors over twenty years and started anew. Within a few years they, and the small group of other workers they had attracted by the simplicity, the precision, and the elegance of their new departure, had made tremendous advances".[17]

Phage genetics thrived in the United States and, after the second world war, was imported into Europe, where it developed rapidly at the Institut Pasteur in Paris. For various reasons it took longer until it became established in Germany.[18] Carsten Bresch, a pioneer of this research in Germany, presents details of its beginnings in his contribution in this volume.

Unlike phage genetics, biochemical and genetic TMV research were imported from the United States in 1937 by Adolf Butenandt, Alfred Kühn and Fritz von Wettstein and established at the Division for Virus Research at the KWIs for Biology and Biochemistry. After the second world war, TMV research was conducted at the MPI for Virus Research, but universities did not take it up. According to botanist and TMV researcher Georg Melchers this was due, among other things, to the "lack of insight into the general importance of a field

[16]Salvador Luria and Max Delbrück, "Mutations of bacteria from virus sensitivity to virus resistance", *Genetics* **28** (1943), pp. 491–511.
[17]Elie Wollman, "Bacterial conjugation", in John Cairns, Gunther S. Stent and James D. Watson (eds.), *Phage and the Origins of Molecular Biology* (New York 1966), p. 216.
[18]For details see U. Deichmann, "Emigration, isolation and the slow start of molecular biology in Germany", *Studies in the History and Philosophy of Biological and Biomedical Sciences* **33** (2002), pp. 433–455.

that was dealing with such a specialised object as the pathogenic agent of a single plant disease" among university professors.[19] In addition, the requirement of a large amount of land and time made the carrying out of TMV research at universities difficult. See also the contribution by Karl-Wolfgang Mundry in this volume.

Phage research, by contrast, had few requirements and gave fast results. Wolfhard Weidel, an early phage and bacteria geneticist at the KWI (MPI) for Biology, explained the general importance of phage research by the fact that it is not "self-contained as may be the case with the systematics of leafhoppers", but "aimed at the solution of biological and biochemical problems of the most general and most far-reaching implications, namely: mode of codification of genetic information in nucleic acid molecules, mechanism of realization of genetic information carried by such polynucleotide chains; mechanism of their exact replication; and, finally, mechanism of mutation and genetic recombination at the molecular level".[20]

The chemists' conservative breakthroughs in genetics: from Friedrich Miescher to Oswald T. Avery

DNA as a phosphorus-containing substance with a high molecular weight was discovered in 1869 by the Swiss biochemist Friedrich Miescher at the University of Tübingen. The "nuclein" which he isolated from nuclei of lymphocytes, consisted — as chemists showed shortly afterwards — predominantly of DNA and some percentage of protein.

For many years the chemical analysis of DNA did not provide any evidence that it possessed the diversity required for the carrier of hereditary information. The assumption formulated in 1906 that DNA

[19]Georg Melchers, "Warum interessiert den Biologen das Tabakmosaikvirus", *Jahrbuch der MPG* (1960), pp. 90–113 (translation by UD). TMV research has been reviewed by Angela N.H. Creager (*The Life of a VIrus. Tobacco Mosaic Virus as an Experimental Model*, 1930–1965, Chicago, 2002); concerning Germany see also Christina Brandt, *Metapher und Experiment. Von der Virusforschung zum genetischen Code*, Göttingen, 2004.

[20]Wolfhard Weidel, "Bacterial viruses (with particular reference to adsorption/penetration)", *Annual Review of Microbiology* **12** (1958), p. 27.

is a small uniform molecule made up of four nucleotides (the tetranucleotide hypothesis) became generally accepted. When the macromolecular nature of DNA was demonstrated in the late 1930s, DNA was thought to be a repetitive polymer comprised of repeating units of tetranucleotides. This hypothesis was critically examined only after Avery's 1944 discovery of DNA's crucial role in bringing about lasting transformations in bacteria (see below).

It should be mentioned that in the 1930s proteins, too, were believed to show regularities in their structures. But due to the large variety of proteins in the cell and the increasing evidence that enzymes and antibodies consist entirely or largely of proteins, the assumption that only proteins are the carriers of biological specificity — that is, the substances responsible for bringing about specific properties of species or individuals and specific biochemical reactions in cells — became almost universally accepted.

When scientists began to experimentally examine the question of the physical and chemical nature of genes in the 1930s, first by radiation studies, and then by virus research, their consensus was that genes, too, must be proteins. The already mentioned crystallization of TMV by Wendell Stanley in 1935 supported this view, because he wrongly identified it as a pure protein. When Norman Pirie and Frederick Bawden showed shortly after that TMV also contained RNA, genes were thought to be nucleoproteins; the specificity lying in the protein part of the molecule.

In the late 1930s a number of different experiments showed the crucial importance of DNA for cell replication and mutation.[21] In 1936 the Swedish biologist Torbjörn Caspersson demonstrated, by UV absorption, that DNA replication took place at the onset of cell division. Several research groups between 1939 and 1941 showed that the UV-mutation spectrum was identical with the DNA absorption spectrum. However, researchers concluded that DNA had only an auxiliary function. The hypothesis that DNA is the material of genes, obvious as it may sound in hindsight, fell victim to the dogma that genes must be proteins. When Erwin Schrödinger, in his well-known book, *What is Life* (1944),

[21]Robert Olby, *The Path to the Double Helix. The Discovery of DNA* (Seattle, 1974), pp. 105–107.

speculated on how genetic information might be stored linearly in genes, he too, took it for granted that genes were proteins.

Oswald Theodore Avery was an outstanding microbiologist and immunochemist at the Rockefeller Institute for Medical Research. In 1944, Avery and his younger associates, Colin MacLeod and Maclyn McCarthy, demonstrated that the substance capable of bringing about a lasting transformation of pneumococcal types — that is, apparently heritable changes — was DNA. This was the first time that a genetic phenomenon was clearly associated to a nucleic acid. It challenged the then generally accepted view that proteins were the material of genes. Their discovery thereby became the basis of all further studies on the structure and genetic functions of DNA.

Avery's paper also had strong methodical impacts. It showed that the physical nature of genes can be analysed *directly* — in contrast to the highly favoured indirect methods of radiation and virus (including phage) research — and it called for chemistry, in particular the chemistry of DNA, to be added as a tool to analyse the gene. As microbiologist Bernard Davis perceived it, "the Avery discovery was truly revolutionary" because of its intrinsic significance and unexpectedness.[22]

Avery's paper was received with open, but restrained, appreciation. However, it was neglected particularly by the phage group around Max Delbrück. As Avery's collaborator René Dubos explained, "certain members of the 'phage group' regarded the orthodox chemical approach to the understanding of biological phenomena as pedestrian, too slow, and not revolutionary enough for their intellectual ambition, ... they did not seem able to do much with or build on [Avery's experiment]".[23] Only when two members of this group, Alfred Hershey and Martha Chase, in 1952 demonstrated the importance of DNA for phage replication, did Avery's conclusion become acceptable to them.

Stimulated by Avery's results, the biochemists Rollin Hotchkiss and Erwin Chargaff in the late 1940s were the first to conduct a quantitative

---

[22]Bernard Davis, *BioEssays* **9** (1988), pp. 130–131. Details on the imapct and reception of Avery's 1944 experiment are in Ute Deichmann, "Early responses to Avery's *et al.*'s 1944 paper on DNA as hereditary material", *Historical Studies in the Physical and Biological Sciences* **34**(2) (2004), pp. 207–233.

[23]Dubos, *The Professor*, p.158.

base analysis of DNA by paper chromatography. Chargaff speculated in 1947 that, among other things, differences in the proportions or in the sequence of the nucleotides forming the nucleotide acid chain could be responsible for specific biological effects. This far reaching hypothesis anticipated linear genetic coding. Two years later he and his collaborators demonstrated that DNA is not a repetitive molecule, but that the DNAs of different species are chemically different. This work led to the discovery of the so-called Chargaff rules of base ratios in DNA (A/T and G/C equal one), first stated in 1950, and which in 1953 found its explanation in the specific base-pairing postulated in Watson's and Crick's double helix theory of DNA structure.

Through their work in X-ray crystallography, physicists and physical chemists had maintained a strong impact on the further development of genetics at the molecular level. Convinced that Avery had shown that genes were made of DNA and not proteins, Maurice Wilkins started X-ray studies on DNA.[24] At a scientific meeting in Naples in the spring of 1951 Wilkins alerted James Watson to X-ray work on DNA. Watson, in turn, motivated Francis Crick in Cambridge, who conducted X-ray studies on proteins for his dissertation, to shift his interest from proteins to DNA.

Outlook — the double helix, the genetic code and biochemistry

The history of the elucidation of the double helix structure of DNA by Watson and Crick in 1953 has been much written about, including by the two scientists themselves. Their work brings together different pathways of, and approaches to, research in genetics. Crick was a physicist who had left physics in order to study the structure of proteins with Max Perutz. Watson was a biologist who, motivated among other things by Schrödinger's book, *What is Life,* had joined the phage group and completed his dissertation under Luria. Watson's

[24]Maurice Wilkins, "DNA at King's College, London", in Donald Chambers (ed.), *DNA: The Double Helix,* Annals of The New York Academy of Sciences, 1995, pp. 200–204.

and Crick's work encompasses X-ray crystallography by Rosalind Franklin and Wilkins, Chargaff's chemical analysis of DNA, and Linus Pauling's theories of the nature of the chemical bond and the role of weak forces such as hydrogen bonds in the structure of macromolecules. They also used Pauling's method of model building.

Max Delbrück pointed to the extraordinary scientific importance of the discovery by comparing it to the discovery of the basic structure of atoms by Ernest Rutherford in his famous gold foil experiment: "Very remarkable things are happening in biology. I think that Jim Watson has made a discovery which may rival that of Rutherford in 1911".[25]

Watson's and Crick's elucidation of the DNA double helix structure opened up another phase of molecular genetics, in which the focus was on gene replication and the genetic code. In their seminal 1953 paper, they made it very clear that they perceived a possible implication of the proposed DNA structure for replication: "It has not escaped our notice that the specific pairing we have postulated immediately suggests a possible copying mechanism for the genetic material".[26] It took four years until the semi-conservative replication mechanism had been solved in principle by Matthew Meselson and Franklin Stahl.

In contrast, the problem of the genetic code was much more complicated. Watson and Crick were the first to use the term "information" for what had been called "biological specificity" before: "The phosphate-sugar backbone of our model is completely regular, but any sequence of the pairs of bases can fit into the structure. It follows that in a long molecule many different permutations are possible and it therefore seems likely that the precise sequence of the bases is the code which carries the genetic information".[27]

In 1948 Shannon had introduced information theory into mathematics. Biologists used the term information increasingly, but usually

---

[25]Max Delbrück to Niels Bohr, 14 April, 1953, Delbrück papers, Caltech Archives.

[26]James D. Watson and Francis H. C. Crick, "A structure for deoxyribose nucleic acid", *Nature 171* (1953), pp. 737–738.

[27]James Watson and Francis Crick, "Genetical implications of the structure of deoxyribonucleic acid", *Nature* **171** (1953), pp. 964–967.

only as a metaphor.[28] Physicists were immediately attracted by the coding problem of DNA. Already in 1954, the astronomer George Gamow, a consultant in the United States Navy during the cold war (thus being well acquainted with coding problems) and, incidentally, a friend of Max Delbrück's, presented an exact phrasing of the coding problem and developed the concept of heredity as information transfer: "The hereditary properties are characterized by a long number written in a four digital system, proteins can be considered as long 'words' based on a 20-letter alphabet. The question arises about the way in which four digital numbers can be translated into such 'words'".[29] Coding was defined as — and largely remained for about seven years — an abstract physical problem independent of biochemical considerations.

Various overlapping triplet codes were theoretically developed by Gamow, Richard Feynman, Edward Teller and other physicists. The arguments for an overlapping code were partly taken from engineering: there would be a better matching of dimensions between protein and DNA-template; storage density would be maximised and waste of information capacity avoided.[30] However, these arguments did not withstand experimental testing. In 1957, Sydney Brenner and Crick ruled out, experimentally and by logical reasoning, an overlapping code. Using frameshift mutations, in which base-analogues (e.g. acridines) were inserted into DNA, they demonstrated for the first time that the code was, as had been predicted, a triplet code. Only three mutations of the same type led to (partially) functioning proteins. Brenner invented the word "codon" to describe the nucleotide unit that would specify an amino acid.

During the next years Leslie Orgel, Carl Woese, Crick and others designed various purely theoretical non-overlapping comma-free codes, which would make sense only if read in one frame. On the

---

[28]Lily Kay, *Who Wrote the Book of Life? A History of the Genetic Code* (Stanford 2000).

[29]George Gamow, "Possible relation between deoxyribonucleic acid and protein structures", *Nature* **173** (1954), p. 318.

[30]Brian Hayes, "The invention of the genetic code", *American Scientist* **86** (1998), pp. 8–14.

assumption of a one-to-one relationship between the number of codons and the number of amino acids, Crick's elegant code coded for just 20 amino acids (among others, the codons AAA, GGG, CCC and UUU were ruled out because, if combined with themselves, they would cause reading-frame ambiguity). Woese later wrote about the fascination with these codes:

> "The comma-free codes received immediate and almost universal acceptance. ... They became the focus of the coding field, simply because of their intellectual elegance and the appeal of their numerology. ... For a period of five years most of the thinking in this area either derived from the comma-free codes or was judged on the basis of compatibility with them."[31]

In 1961 two outsiders of the field experimentally refuted Crick's elegant proposal and deciphered the first "code word". Heinrich Matthaei and Marshall Nirenberg showed that artificial RNA can stimulate protein synthesis in a cell-free system: poly-Uracil RNA coded for poly-phenylalanine, the codon for this amino-acid thus being UUU. Their work opened up a next, largely biochemical phase in the history of the genetic code, in which the codons for all amino acids were deciphered in about six years. In the end, the code did not resemble most of the theoretical notions: the magic number of 20 does not play a role; all the mathematical solutions to solve the frame-shift problems are not used; and the code is redundant (many amino acids have several codons each).

Is it thus fair to say that physicists were mainly wrong and biochemists mainly right? Such a simplistic assessment completely neglects the decisive role of theory. As Carl Woese later observed, "What has not generally been appreciated is that the subsequent spectacular advances occurring in the second period [1961–67] were interpreted and assimilated with ease, their values appreciated, and new experiments readily designed precisely because of the conceptual framework that had already been laid".[32]

---

[31]Cited after Hayes, ibid.

[32]Kay, *Who Wrote the Book*, p. 129.

Partly parallel to the developments discussed here, decisive contributions were made to the elucidation of protein biosynthesis (see also the contribution by Hans Zachau in this volume) and gene regulation. For reasons of brevity, they are not discussed here. The articles collected in a recent volume highlight the importance of biochemistry in studies on protein biosynthesis.[33]

In a paper titled "Biochemistry strikes back", Sydney Brenner recalls the changing relationship between theoretical approaches and (biochemical) experiments in molecular biology: "The early days of molecular biology were marked by what seemed to many to be an arrogant cleavage of the new science from biochemistry".[34] But molecular biology, according to Brenner "a way of life and not a subject", effected a fusion between biochemistry and genetics. He holds the conviction that despite the recent increase and success of "omic science" — genomics, transcriptomics, proteomics — problems such as protein interactions will not be solved by proteomics. Biochemistry will flourish in the future, too, "because it provides the only experimental basis for causal understanding of biological mechanisms."[35] Jacques Loeb's programme still appears to be alive. Many contributions to this volume bear witness to it.

---

[33]Jan Witkowski (ed.), *The Inside Story. DNA to RNA to Protein* (New York 2005).

[34]Sydney Brenner "Biochemistry strikes back", in Jan Witkowski (ed.), *The Inside Story*, pp. 367–369.

[35]Ibid.

II. First Initiatives, Concept, Founding and Crisis

2

Founding and Crisis

Simone Wenkel

On 22 June 1962, the Institute of Genetics in Cologne (hereafter Institute) was inaugurated. Following a plenary lecture by the physicist Niels Bohr, the first institute dedicated exclusively to molecular biological research was opened at a German university. At that time the plans for the new institute were almost ten years old. Facing great commitment, generous support and despite some difficulties, the Institute had an unusual start in 1961. The following listing of faculty of the Institute between 1956 and 1972 demonstrates clearly the title of this contribution: Founding and Crisis.

1956	Delbrück
1958	Bresch
1960	Bresch, Harm
1961	Bresch, Delbrück, Harm, Starlinger, Zachau
1962	Bresch, Delbrück, Harm, Henning, Starlinger, Zachau
1963	Bresch, Harm, Henning, Starlinger, Zachau
1964	Starlinger, Henning, Zachau
1965	Starlinger
1967	Starlinger, Vielmetter
1968	Müller-Hill, Starlinger, Vielmetter
1970	Müller-Hill, Rajewsky, Starlinger, Vielmetter
1972	Doerfler, Müller-Hill, Rajewsky, Starlinger, Vielmetter

The founding of the Institute took place during the rebuilding of Germany after the second world war, in the phase of reconstruction of German academic institutions. The unique situation for inaugurating molecular biological research in Germany derived from the fact that it had to be imported completely in the 1950s and 1960s.[36] In France, the United Kingdom, and the United States molecular biology made major advances from the late 1940s and early 1950s, however, German research was not yet involved in this endeavour. Molecular biology, as it is understood in this article, includes research in bacterial and phage genetics and research on the structure and function of genetic material and proteins. Except for two groups conducting research in biochemistry and genetics of the tobacco mosaic virus in Tübingen under Georg Melchers and Gerhard Schramm, and two small phage groups in Göttingen (Carsten Bresch) and Tübingen (Wolfhard Weidel) — all of them located at Max Planck Institutes — there existed no molecular biological research groups. Furthermore, there was no teaching of molecular biology and hardly any of biochemistry and biophysics in Germany before the mid-1950s.

The idea

In 1953 Joseph Straub, the head of the Botanical Institute at the University of Cologne, was offered a position at the University of Munich. During the negotiations for this position he wrote to the Dean of the Faculty of Natural Sciences in Munich:

> "The development of biology is taking place more and more into the direction of microbiology, and in this process the genetically oriented research should be the most significant. ... I do not know any biological university institute in Germany, whose research is dedicated to this field.

[36]The international history of molecular biology has been discussed in numerous publications, for example: Michel Morange *A History of Molecular Biology* (Cambridge 1998); Bruno J. Strasser and Soraya de Chadarevian *(eds.)*, "Molecular Biology in Post-War Europe" special issue of *Studies in the History and Philosophy of Biological and Biomedical Sciences*, vol 33C (2002); Horace F. Judson, *The Eighth Day of Creation* (New York, 1979).

... I am convinced that the most valuable contributions to the development of the biological sciences in Germany will be achieved by such an institute, whose work will focus on the genetics of micro-organisms."[37]

Though Munich did not support Straub's plans sufficiently, the call there led Cologne to create an additional post of associate professor [*Extraordinariat*] of microbiology at the Botanical Institute. In order to find an appropriate person for this new position, Straub contacted Max Delbrück, an émigré physicist and founder of phage genetics, for suggestions. Remarkably, not long after their first contact this associate professorship was offered to Delbrück himself. This created an absurd situation: an associate professorship at a botanical institute in Germany was offered to a long-established, world-leading scientist in molecular biology. Delbrück, naturally, did not even consider the offer, rejecting it with a few words on a postcard, which did not endear him to the administration.[38] As compensation, however, Delbrück offered to visit Cologne as guest professor, which was accepted by Straub. During Delbrück's stay in Cologne, he and Straub made the first plans to found a molecular biological institute at the university.

The founders

For co-founders of an institute, Max Delbrück and Joseph Straub had different backgrounds and motivations. Straub (1911–87; Fig. 1) studied sciences and finished his dissertation and *Habilitation* with the botanist Friedrich Oehlkers in Freiburg; later he worked at the Kaiser Wilhelm Institute for Biology in the department of Fritz von Wettstein. In June 1949 Straub was appointed full professor of botany at the University of Cologne. He had a strong interest in teaching genetics and, due to his contact and discussions with Georg Melchers in

[37]Joseph Straub to E. Wiberg (dean of the faculty of natural sciences), University of Munich, Sept. 28, 1953, Archive of the Max Planck Society, Berlin, Abt. III, Rep 75/6 (translation SW).

[38]Max Delbrück to Oberregierungsrat von Medem, July 6, 1955, Universitätsarchiv Köln, Zug. 343 Nr. 2.

Fig. 1. Joseph Straub.

Tübingen and Hans Marquardt in Freiburg, he closely followed developments in the "new biology". Straub did not want to conduct genetic research himself, but was interested in bringing it to his institute. He had proved his diplomatic skills during negotiations in the faculty, which led to the separation from the philosophical faculty of what became a new faculty of mathematics and sciences in the early 1950s. He had no fear of building up an institute that topped his own, both in design and manpower. In 1961 Straub became director of a department at the Max Planck Institute of Plant Breeding Research in Cologne-Vogelsang, while remaining university professor at Cologne.

Max Delbrück's (Fig. 2) biography illustrates his background intentions in the field of molecular biology and in German research.[39] Delbrück, a theoretical physicist who emigrated to the United States in

[39]See also the introductory chapter in this volume.

Fig. 2. Max Delbrück, 1972.

1937 with a scholarship from the Rockefeller Foundation, kept strong ties to Germany. In the United States he was co-founder — with Salvador Luria — of phage genetics and father of the "phage group" which pioneered this research for many years. His laboratory at the California Institute of Technology (Caltech) was a training centre for many students and scientists who later became molecular biologists, among them Seymour Benzer, Gunther Stent and James Watson. His interest in teaching is shown in the initiation of the Cold Spring Harbor phage courses, again together with Luria, in the mid-1940s. These courses provided opportunities for many graduated scientists to learn phage genetics.

When Delbrück returned to Germany for the first time after the second world war, in 1947, he started efforts to bring molecular biology to Germany. In his laboratory at Caltech many young German post-doctoral fellows learned phage genetics in the 1950s and 1960s. He played a leading role in the establishment and institutionalisation of molecular biology in Germany, particularly in Cologne and Konstanz, being a regular guest professor at both universities until his death in 1981. In 1969 Delbrück was awarded the Nobel Prize for Physiology and Medicine, together with Alfred Hershey and Salvador Luria, for their discoveries concerning the replication mechanism and the genetic structure of viruses. Although Delbrück until the 1960s encouraged scientists to conduct phage genetics, he focussed his own research on the analysis of light reactions of the fungus *Phycomyces* from the 1950s onwards. This work remained, however, without major success.

The foundation

Delbrück had a vision that the new institute should be designed as a model for other German universities. He wanted to introduce his favoured departmental system of many independent but interacting research groups to the German university structure. It had proved highly fruitful at Caltech. An arrangement such as this one would, however, only be possible if he built up an all-new institute out of a "vacuum",[40] as he called it. His project was far more than just a research institute, as demonstrated by the fact that he rejected an offer from the Max Planck Society to come to Tübingen as founding director of a Max Planck Institute for Virus Research.[41] He wrote to his brother-in-law Karl Friedrich Bonhoeffer that:

> "What appeals to me with the project in Cologne, is not the size of the institute, the salary or the budget, but the chance to actively break down the organisational deadlock in academic biology, together with Straub, in whose insight, skills and energy I trust very much. I think if we succeed, German biology will be helped ten times more than by a small research institute on a hill outside of Tübingen, to which no

[40]Max Delbrück to George Beadle, 6 September, 1956, Caltech Archives, Delbrück papers, Box 2.10.

[41]Max Delbrück to K.F. Bonhoeffer, 24 November, 1952, Caltech Archives, Delbrück papers, Folder 4.5.

student will ever stray, and whose influence on academia the established interests of the faculty would resist unanimously."[42]

Delbrück also explained his aims in his 1964 account of the Cologne Institute's development:

"The liberal atmosphere at the Institute is caused by the many group leaders of the same status. As a consequence, the group leaders interact very often and necessarily disagree with one another, so all other members of the Institute learn to discuss more freely than it would be possible at institutes where there is only one director. Being often the most experienced person at the institute, he easily gets used to consider himself a person without flaws."[43]

Nevertheless, Delbrück's idea was not so much attached to Cologne, as it might seem to be, but more to the abilities of Joseph Straub. Maybe the Institute would have been in Munich if Straub had agreed to move there. It is also important to mention the other agencies involved, notably, the leading characters of the main German funding agency, "*Deutsche Forschungsgemeinschaft*" (DFG), the university and the Ministry of Sciences in North Rhine-Westphalia, all of which overwhelmingly supported Straub's plans.

After an informal meeting with Delbrück and these agencies in Bad Godesberg Straub submitted a grant application to the DFG on 20 November 1957, and applied for funds to build an Institute of Genetics at the University of Cologne. Altogether he raised DM 2.5 million. In the 11-page application he described his project and pointed out that support for modern biology, which he called "gene physiology", in Germany was at that time non-existent.

"The situation is so serious that one can say: if Genetics in Germany will not receive a crucial and reasonable promotion, we will become

[42]Max Delbrück to K. F. Bonhoeffer 11 November, 1956, Caltech Archives, Delbrück Papers, Folder 4.5, (translation by SW).

[43]Max Delbrück wrote his account on the history of the Institute of Genetics in Cologne in a five paged summary in 1964, which was circulated in the Institute. (Copy at the archive of the Institute of Genetics in Cologne.)

an underdeveloped nation [*volk*] in biology, because the theoretical and practical education in this biological discipline does not keep up with the state of research abroad."[44]

In the description of the project, he pointed to the former leading role of Germany in genetic research, and its subsequent decline, and emphasised that genetics with microorganisms was already founded and fast developing in the United States and that Germany was in danger of falling far behind. Straub emphasised the importance of teaching the new biology, which, according to him, had been impossible until then. He argued that no well-trained scientists were available and university training in biology was not arranged in an interdisciplinary way, drawing attention to the fact that too little physics and chemistry were taught to biology students. Straub proposed a solution for these problems: a big institute with autonomous, interdisciplinary groups to support collaboration and multifaceted teaching and research in genetics of microorganisms. He highlighted the excellent conditions in Cologne, where renowned scientist Max Delbrück would lead the Institute.

The application was approved in 1958. In the same year Carsten Bresch and his group started work at the Botanical Institute, where they stayed until they moved into the new building at the end of 1960 (Fig. 3). (Between 1958 and 1960 the future members of the new Institute located themselves at the "Institute of Genetics under construction".)

The awareness of the importance of a university institute for molecular biology and the opportunity of having Max Delbrück as its director led to very generous financial support. Since the DFG was not allowed to provide building funds out of its own budget, the approved application was forwarded to the *Wissenschaftsrat*, as deciding board of the Federal Government of Germany. The *Wissenschaftsrat* approved DM 1.6 million from Federal funds. The remaining costs would be paid by the State of North Rhine-Westphalia. The total costs for the building were DM 3.39 million,

---

[44]Joseph Straub to the DFG, 20 November, 1957, Caltech Archives, Delbrück papers, Box 21.5 (emphasis in the original, translation by SW).

Fig. 3. The pictures show the building of the Institute of Genetics between 1959–1960.

including DM 1.79 million paid by the Federal State of North Rhine-Westphalia.[45] The Institute was equipped for DM 250,000 in 1960 only,[46] and the total amount for the equipment was DM

[45]Freiherr von Medem, 29 October, 1958, Hauptstaatsarchiv of North Rhine-Westphalia, NW 144–887. The amount paid by the Federal State of North Rhine-Westphalia includes DM 420,000 allowed in 1961 to finish the building (Kultursministerium, 7 February, 1961. Hauptstaatsarchiv of North Rhine-Westphalia, NW 144–887).

[46] Wagner to Max Delbrück, 15 December, 1959. Caltech Archives, Delbrück papers, Box 26.22.

Fig. 4. Opening ceremony at the Institute Max Delbrück (*right*) Niels Bohr (sitting at right in the front row).

2.5 million to fully furnish all departments.[47] An additional half million DM was provided to Max Delbrück in 1963 by the newly founded *Volkswagenstiftung* to equip the biophysical group of Ulf Henning.

At the opening ceremony in June 1962 Niels Bohr, Delbrück's mentor, held the lecture "Light and life revisited" (Fig. 4).[48] Delbrück headed the Institute from 1961 to 1963.

The early years

The early years of the Institute were characterised by some features that differed markedly from other German biological institutes; for example, the informal atmosphere, the existence of young independent group leaders and foreign guests (Fig. 5). These characteristics are dealt with in many other contributions in this volume. Another

---

[47]"Zur Geschichte des Institutes für Genetik der Universität zu Köln" (1964) p. 2; Folder "Baufragen" at the Archive of the Institute of Genetics, Cologne. The same folder contains a detailed cost estimate for the Institute's equipment calculated in 1959 adding up to about DM 1.63 million.

[48]See the reprinted article of Niels Bohr in this volume.

Fig. 5. A seminar at the Institute about 1963. Peter Starlinger (*left*) and Thomas Trautner (*right*).

peculiarity was that almost all group leaders were not classical biologists, that is, botanists or zoologists, but chemists, physicists and microbiologists who had received most of their molecular biological training in the United States, the United Kingdom or in France (see Table 1). Notably, Carsten Bresch was not permitted to go to the United States to work with Max Delbrück because of his leftist political background. Instead he received material from Delbrück, with whom he also discussed his results in many letters from the late 1940s onwards.

Delbrück established a phage course in Germany that was modelled after the Cold Spring Harbor Phage Course which he had initiated (Fig. 6). These phage and bacteria courses in Cologne were the most important reasons for the popularity of the Institute of Genetics. Many of the participants later became professors, and the Institute had a chance to lure ambitious young scientists to Cologne for extended stays. The tradition of the phage courses has lasted until today as the annual "Cologne Spring Meeting".

Table 1 Group leaders at the Institute between 1958 and 1972, and the institutions outside Germany where they received at least part of their education in molecular biology prior to joining Cologne (in brackets: supervisors).

Walter Harm	California Institute of Technology, USA (Max Delbrück)
Peter Starlinger	California Institute of Technology, USA (Max Delbrück)
Hans Georg Zachau	Massachusetts Institute of Technology, USA (J.C. Sheehan); Rockefeller Institute, USA (Fritz Lipmann)
Ulf Henning	Stanford University, USA (Charles Yanofsky)
Walter Vielmetter	Cold Spring Harbor Laboratory, USA; University of Cambridge, UK
Benno Müller-Hill	Indiana University, USA (Howard Rickenberg); Harvard University, USA (Walter Gilbert and James D. Watson)
Klaus Rajewsky	Institut Pasteur, Paris, France (Pierre Grabar)
Walter Doerfler	Stanford University, USA (David Hogness); Rockefeller University, USA

The crisis

Max Delbrück's plan for the local situation was very optimistic. He wanted the Institute built with no disadvantage to the other biological institutes. He proposed direct physical connections between the institutes' buildings, shared facilities and a "MONTAN-UNION"[49] with much collaboration.

But reality was less ideal. The unique characteristics and what were perceived to be certain financial privileges of the Institute of Genetics led to problems at the university. The newly founded Faculty for Mathematics and Sciences in Cologne was still busy rebuilding all the established institutes after the second world war. Despite strong external support, a new institute of the size planned held a certain risk of discrimination at the expense of other institutes. Although most members of the faculty supported Straub's plans, criticism grew after faculty members observed apparent disadvantages for their own institutes. The ensuing resentments probably prevented some fruitful collaboration between the biological institutes. Together

[49]Max Delbrück to Wolfram Zillig, 18 April, 1960, Caltech Archives, Delbrück Papers Folder 25.16 (emphasis in the original). The Montan-Union was the European Union of Coal and Steel, founded in 1951.

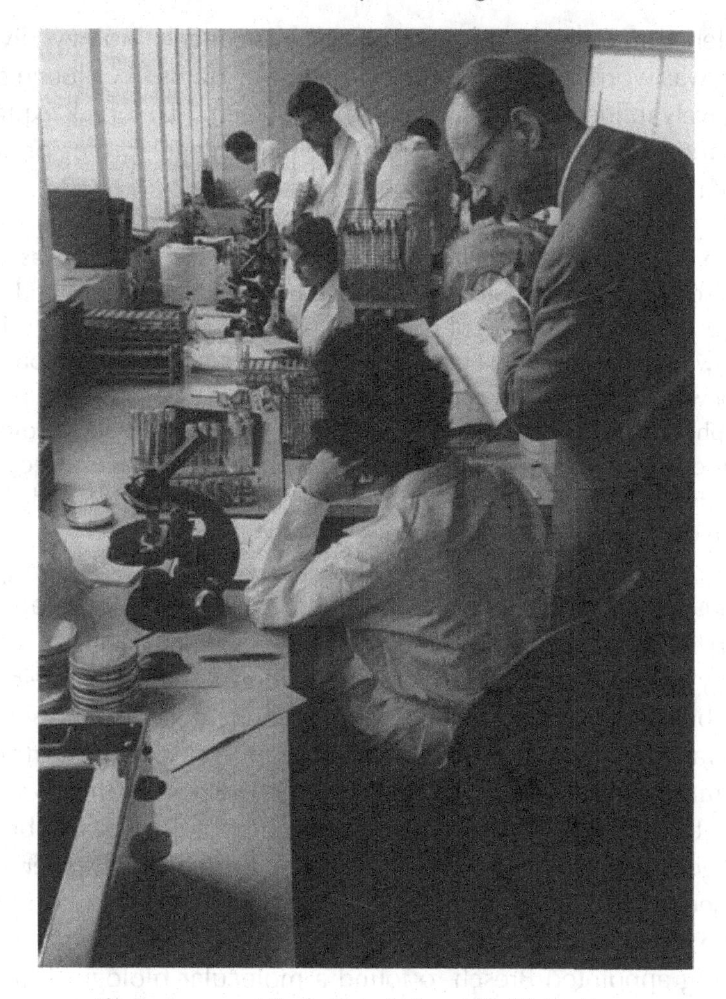

Fig. 6. A course at the Institute, in the middle Walter Harm.

with the strong international orientation and the international recognition of the Institute, as well as the prejudices of some biologists against the new style of research, these local problems most probably contributed to the crisis following Delbrück's departure in 1963 — he remained director "on leave".

The faculty did not appoint Carsten Bresch and Walter Harm as co-directors and refused to establish the planned "polycephalic" principal for the institute. The uncertain status of Delbrück also caused problems at Caltech, where Ray Owen, Caltech biology division chair, was worried that Delbrück's "on leave" status in Cologne might negatively influence Delbrück's work at Caltech.[50] Delbrück expressed his impressions in a letter to Bernhard Mühlschlegel, professor of theoretical physics in Cologne:

> "On my departure, we achieved only a compromise. My idea was that Bresch and Harm should immediately become full professors and co-directors of the Institute of Genetics and that I can completely withdraw from the Institute. Some members of the faculty commission did not want to do that, especially they did not want to establish the polycephalic principal. The compromise was that I officially stayed director on an "on leave" status and that Bresch and Harm become co-directors provisionally.... .The only possibility for me is to radically rule out a returning to the Institute."[51]

Thus Delbrück did not intend to come back and almost all other group leaders left the Institute between 1964 and 1967. They left mainly because of the provisional situation concerning their positions but also because an offer for a position as head of a big institute was more attractive than being a group leader in a departmental system. Having received international recognition for their scientific work, the shortage of well-trained molecular biologists also brought very good offers from overseas. Carsten Bresch and Walter Harm decided to accept offers from the University of Dallas, in Texas. Bresch left Dallas for Freiburg only three years later. The University of Freiburg appointed Bresch to found a molecular biological institute with a similar structure to that in Cologne.[52]

Ulf Henning was appointed director of a department at the Max Planck Institute for Biology in Tübingen. He was soon followed by Peter Overath, who, after finishing his *Habilitation* in Cologne, also became director at the same institute in Tübingen. Hans Georg

---

[50]Ray Owen to Max Delbrück, Caltech Archives, Delbrück papers, Box 17.7.

[51]Max Delbrück to Bernhard Mühlschlegel, 1963, Caltech Archives, Delbrück papers Box 16.19.

[52]See Carsten Bresch's contribution in this volume.

Zachau accepted an offer as full professor of physiological chemistry at the University of Munich in 1966, and left Cologne in 1967. Peter Starlinger received two offers, one as director of the newly founded Max Planck Institute of Molecular Genetics in Berlin, and the other one from the University of Bochum. But Starlinger decided to stay in Cologne, where he eventually was appointed director at the Institute of Genetics in 1965. Through his decision to stay in Cologne as the first permanent director of the Institute, Starlinger prevented its closure.

Between the years 1967 and 1972 four additional co-directors were appointed full professors at the Institute. The first one, Walter Vielmetter, a biologist and phage geneticist, had worked with Hans Friedrich-Freksa and Heinz Schuster on phage mutations at the Max Planck Institute for Virus Research in Tübingen. Benno Müller-Hill, a chemist, was appointed in 1968 without *Habilitation.* He undertook his Ph.D. in Freiburg with Kurt Wallenfels. Working on the genetics of the lac system at Harvard University, he had isolated the lac repressor together with Walter Gilbert in 1966. Klaus Rajewsky studied medicine in Munich and dedicated his research to immunology. He worked at the Institute in Cologne for several years before finishing his *Habilitation* and was appointed full professor in 1970. The last co-director, appointed in 1972, was Walter Doerfler, who had also studied medicine. He was originally a virologist working with Wolfram Zillig at the Max Planck Institute for Biochemistry in Munich and prior to his appointment in Cologne was assistant professor at the Rockefeller University.

The research

Early on the Institute's research received high international recognition.[53] The first research groups starting in 1958 followed Max Delbrück's guidance and focussed their work on phage genetics. In 1960 Delbrück successfully persuaded the biochemist Hans Georg Zachau to come to Cologne and lead a group. Among the highly

[53]A detailed analysis of the research conducted at the Institute will be published in my forthcoming dissertation.

cited papers until the early 1970s were the elucidation of the secondary structure of transfer RNA by Zachau and his group and the discovery of insertion elements in bacteria by Peter Starlinger and his group.[54] Work by Walter Harm on the UV sensitivity of T4 phages; Peter Overath on the mechanisms involved in fatty acid degradation; and Konrad Beyreuther in Benno Müller-Hill's group on the function of lac repressor, led to a high international profile for the Institute.[55]

The number of publications from the Institute and the degree to which they were cited (see Tables 2 and 3) reflected the different phases of its early history. It should be mentioned that most publications by the Institute's members, even if published in German journals, were written in English; a fact that was most unusual for German institutions, where the language used in publications did not change from German to English until the late 1960s.

The number of citations per publication can be used to assess the impact of scientific research (see Table 3).[56] The figure shows the increasing recognition of the Institute's members' publications from the early 1960s onwards and the importance of the results obtained at the Institute between 1968 and 1972. The extraordinarily high number of citations per publication in 1970, the year in which only three papers were published, reflects the impact of Peter Overath's paper on fatty acid degradation in *E. coli* (372 citations).

---

[54]Hans Georg Zachau, Dieter Dütting and Horst Feldmann, "The structures of two serine transfer ribonucleic acids", *Hoppe Seylers Zeitschrift für Physiologische Chemie 347* (4), (1966), pp. 212–235; Elke Jordan, Heinz Saedler and Peter Starlinger, "0^0 and strong-polar mutations in the gal operon are insertions", *Molecular and General Genetics* 102 (1968), pp. 353–363.

[55]Walter Harm, "Mutants of phage T4 with increased sensitivity to Ultraviolet", *Virology 19* (1963), pp. 66–71; K. Adler, K. Beyreuther, E. Fanning *et al.*, "How lac repressor binds to DNA," *Nature* **237** (1972), pp. 322–327; Peter Overath, Georg Pauli, Hans U. Schairer, "Fatty acid degradation in Escherichia coli. An inducible Acetyl-CoA Synthetase, the mapping of old-mutations and the isolation of regulatory mutants", *European Journal of Biochemistry* **7** (1969), pp. 559–574.

[56]Citations shown in Fig. 2 data originate from ISI Web of Knowledge [Version 3.0; URL: http://isiknowledge.com]; April 2005. Details about this method and its problems will appear in my forthcoming dissertation.

Table 2 The number of publications per year by the molecular biological groups at the Institute between 1958 and 1978. The impact of the crisis can be seen in the decline of publications in the early 1970s which, due to the delay in publishing experimental data, points to a crisis in the late 1960s.

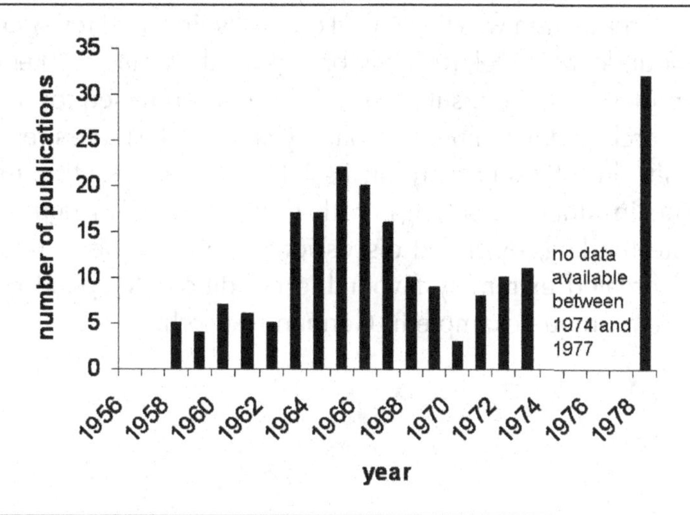

Table 3 The average number of citations per publication at the Institute between 1958 and 1978.

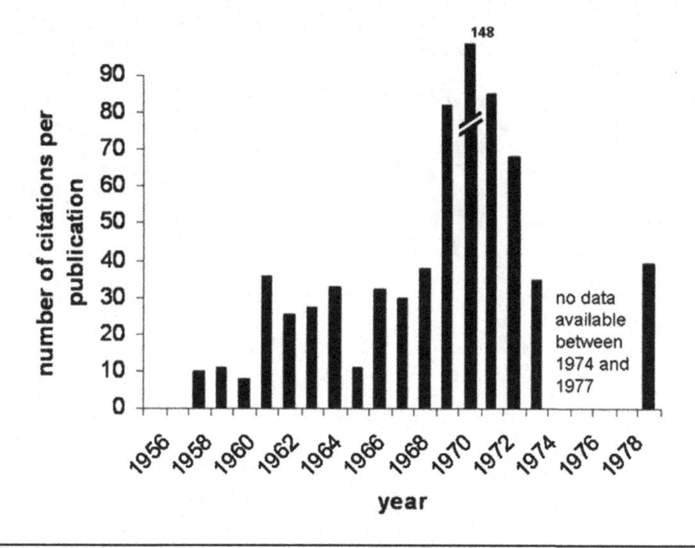

Conclusion

The Institute of Genetics at the University of Cologne was one of the leading institutes of molecular biology in Germany in the 1960s and 1970s. Its reputation was first based on the "spirit" and research focus that its founder Max Delbrück had brought to the Institute. After a crisis in the mid-1960s, the Institute soon became renowned for its excellent research and teaching. For many years, the Institute's reputation was built almost exclusively on its international scientific recognition. The institutional settings, with group leaders of equal status, adequate funding, extended discussions and teaching in the phage courses created extremely favourable conditions for good research and were a unique example in German research.

3

Die erste Zeit

Carsten Bresch

1945–49

Am Ende des Krieges lag alle Wissenschaft in Deutschland am Boden. Moderne Biologie gab es wohl nur in Heidelberg und vor allem in Tübingen um Butenandt und um die Virus-Gruppe mit Georg Melchers, Gerhard Schramm, Hans Friedrich-Freksa, Werner Schäfer, Anton Lang und Wolfhard Weidel, dem ersten modernen Phagenmann in Deutschland.

Die Zeit war hoch politisch — besonders betroffen was die Genetik — speziell die Humangenetik, eng verstrickt in die Verbrechen des Nationalsozialismus. Bis dahin führende Wissenschaftler wie Otmar v. Verschuer oder Fritz Lenz kämpften mit „Persilscheinen" um frühere Positionen — bestens nachzulesen bei Benno Müller-Hill. (Übrigens: Fritz Lenz — bitte nicht zu verwechseln mit seinem verdienstvollen Sohn Widukind Lenz, der den Zusammenhang zwischen Contergan und Mißbildungen aufdeckte.)

Die Zeit war hochpolitisch — auch für Unbelastete oder Nazigegner — ganz besonders in Berlin. Man lebte im Zerreißen der Welt in Ost und West. Es war die Phase des kalten Krieges. So war Prof. Hans Nachtsheim, obgleich Abteilungsleiter am Kaiser Wilhelm-Institut für „Anthropologie, menschliche Erblehre und Eugenik" einer der wenigen unbelasteten Genetiker. Er behielt seine Professur an der Humboldt-Universität in Ostberlin und plante sogar auf Einladung der sowjetischen Behörden den Umzug seiner KWI-Abteilung von Dahlem im US-Sektor nach Berlin-Buch, und zwar in das Institut, in dem einst Max Delbrück mit Timoféeff-Ressovsky gearbeitet hatte und das heute „Max Delbrück Centrum" heißt. (Timoféeff wurde schon 1945 verhaftet und nach Rußland gebracht.) Bevor Nachtsheim umgezogen war, verschärfte sich die Ost-West-Spannung zusehends: am 19. Juni 1948 begann die Blockade Westberlins und etwa zur gleichen Zeit wurde die „Freie Universität" im US-Sektor gegründet. Nachtsheim blieb im Westen.

Ich selbst war 1941 mit einem Physik-Vorexamen Soldat geworden und nahm mein Studium im Sommer 1946 an der Humboldt-Uni wieder auf. Ich wohnte in Westberlin, war aber politisch fest im Osten engagiert. Das änderte der Bericht zur Tagung der Lenin-Akademie im Sommer 1948. Diese höchstpolitische Tagung unter dem Titel „Die Lage in der biologischen Wissenschaft" sollte den sogenannten „Mendelismus-Morganismus-Weismannismus" vernichten und durch die Pseudowissenschaft von Lyssenko ersetzen. Tatsächlich war es die menschenverachtende Entwürdigung der alten russischen Genetiker. Erst im April 1956 beendete der Druck der molekular werdenden Genetik auch in der Sowjetunion diese Schande.

Max Delbrück hatte gleich nach Kriegsende seine familiären und wissenschaftlichen Bindungen in Deutschland erneuert. Dazu gehörten zwei Vorträge im Berliner Harnackhaus. Vom ersten, am 21. August 1947, berichtete mir Wolfgang Eckart, mein langjähriger Studienkollege und leider viel zu früh verstorbener Freund. Er war durch Delbrücks Vortrag total Phagen-infiziert und übertrug seine Begeisterung sogleich auf mich. Wir erkundeten schnell in Ost- und West-Berlin (damals konnte man sich noch freizügig bewegen), in welchem Institut mit Phagen gearbeitet wurde. Tatsächlich gab es ein einziges: das Robert-Koch-Institut in Wedding. Das war Westberlin.

Der Chef des Phagen-Labors und zugleich Direktor des ganzen Institutes war ein äußerst liebenswürdiger alter Herr – der Geheime Obermedizinalrat Prof. Dr. Otto Lentz (Lentz mit tz, also keine Verwandtschaft zum vorerwähnten Humangenetiker Lenz). Offenbar beeindruckt von unserem Phagen-Enthusiasmus nahm er uns bereitwillig in sein Labor auf, schickte uns aber erst einmal ins Pestlabor,

„um wirklich steriles Arbeiten zu lernen", wie er sagte. Wir rätselten, ob es nicht vielleicht wirklich *Pasteurella pestis* war, was wir da überimpften und kultivieren mußten? Obwohl Robert Koch der Erfinder der festen Nährböden war und D´Hérelle auch schon Phagen-Löcher (taches) im Bakterienrasen beschrieb, wurde der Phagen-Titer noch durch das Aufklaren einer Verdünnungsreihe ermittelt.

Ich hätte Ihnen heute erzählt, daß uns Delbrück die T-Phagen für unsere Arbeit geschickt hätte, wäre da nicht in meinen Akten der Durchschlag eines Briefes an Delbrück vom Oktober 1948, aus dem hervorgeht, daß wir Delbrücks zwei Jahre alte Phagen-Sendung schon aus der Schreibtischschublade von Prof. Robert Rompe erhalten hatten. Rompe hatte den für uns zuständigen Physik-Lehrstuhl an der Ost-Universität. Zu unserer Freude stimmte er zu, den experimentellen Teil einer Doktorarbeit mit selbst gewählter Fragestellung im westlichen Robert-Koch-Institut zu machen. Damit waren wir also beschäftigt, als die russische Blockade begann, die im Westsektor permanente Stromsperren nötig machte. Im Institut gab es Strom nachts von 3 bis 4 oder 5 Uhr. So fuhr ich jede Nacht mit dem Fahrrad eine Stunde durch die dunkle Stadt und nach gelungenem Experiment im Morgengrauen ebenso zurück.

Mit Delbrück standen wir in ständigem Briefkontakt. Er schickte uns in wirklich väterlicher Fürsorge seine Sonderdrucke und auch die seiner Kollegen, die er für uns in der Vor-Xerox-Welt besorgte. Diese ständige Hilfsbereitschaft — gerade den wissenschaftlichen Anfängern gegenüber — war einer der herausragendsten Verdienste von Max Delbrück.

Delbrücks Fürsprache war es auch, die mir nach bestandener Promotion eine Stelle in Göttingen bei Prof. Karl-Friedrich Bonhoeffer verschaffte. Sein Institut, das MPI für Physikalische Chemie, zeichnete sich aus durch ein wissenschaftlich, politsch und menschlich gleich wohltuendes Klima.

1950–55

Die Welt der Virologen war seit dem Kopenhagener Polio-Kongreß im September 1951 fasziniert von Delbrücks Vortrag zur „mating-oder

Visconti-Delbrück-theory". Diese betrachtete die mehrfach infizierte Wirtszelle als so etwas wie einen großen Mäusekäfig, in dem zwischen verschiedenen Phagenmutanten zufällige konsekutive Paarungen mit genetischer Rekombination stattfanden. Die schließlich freigesetzten Phagen konnten so Gene von sogar mehr als zwei Eltern tragen.

Die quantitative Behandlung der Theorie war einigermaßen komplex. „Biologen beklagen sich", schrieb Delbrück in einem Brief, „weil sie nicht gewohnt sind auf Arbeiten zu treffen, die sie nicht wie eine Zeitung lesen können". Als Delbrück gefragt wurde, wofür die Abkürzung „v" in der Theorie stünde, sagte er in tiefem Ernst: "I called it „v" for Visconti". Visconti war ein Gastprofessor in Delbrücks Lab, wobei niemand recht entscheiden wollte, ob er mehr an Phagen-Genetik oder an Delbrücks Frau Manny interessiert war.

Ich hatte das Glück, Delbrücks Vortrag in Kopenhagen miterleben zu können. Zur Kongreß-Einladung hatte mir Günther Stent verholfen, der hier wie oft auch später mir ein guter Freund und Helfer war. Ich erlebte das erste Mal die Diskussionen im kleinen Kreis: Max Delbrück, Günther Stent, Jim Watson, die Kopenhagener Niels Jerne und Ole Maaloe und auch dabei Niels Bohr — für einen jungen Physiker ein Halbgott der Wissenschaftsgeschichte — dessen Pfeife dauernd ausging, sodaß er meine ganze Schachtel Streichhölzer verbrauchte (welcher Unsinn so haften bleibt).

Die Phagenwelt um Delbrück und Luria war ein großer Sauerteig-eine große Gemeinschaft, in der man sich ständig zu Seminaren einlud oder bei kleinen Konferenzen mit vielen Partys traf, in langen Briefen nicht nur die eigenen Ergebnisse und Pläne sondern auch die der Kollegen offen diskutierte — meist mit einer Portion Tratsch garniert. Es gab ein inoffizielles hektographiertes Nachrichtenblatt, den "Phage Information Service", der durch Delbrück an alle Freunde verteilt wurde. Lange vor der offiziellen Veröffentlichung kannte so die in-group oft schon das Kommende. Noch dachte niemand an Patent-Anmeldungen — noch stritt niemand um Prioritäten. Man wollte nur wissen, wie die Natur funktionierte.

Im Sommer 52 — es war ein Jahr nach Kopenhagen — fand die Tagung in Royaumont bei Paris statt — im Jahr darauf kam die Sensation der Doppelhelix (*Nature* 25.4. und 30.5.1953). Delbrück war schon vorab durch einen langen Brief von Jim Watson informiert — als Faksimile nachzulesen in Watsons Buch. Delbrück schrieb mir im Juni, nun müßten wohl alle Fakten der Genetik neu interpretiert werden. Später formulierte er den Satz: „Es wäre kaum glaublich, daß die Natur von der wunderbaren Erfindung der Herren Watson und Crick keinen Gebrauch gemacht hätte".

Für mich war der Höhepunkt dieser so dynamischen Zeit ein 3-Monate-Aufenthalt der ganzen Familie Delbrück, d.h. mit Manny und den Kindern Jonathan und Nicola (wohl sieben und fünf Jahre alt) am Bonhoeffer-Institut in Göttingen. Aus Prof. Delbrück wurde damals „Max" und Max arbeitete im Labor nebenan — stets bereit über Phagen zu reden, aber nur zu reden. Der Wermutstropfen war *Phycomyces*, Max´s neue große Liebe, um die sich alle seine eigenen Versuche drehten. Hatte Max das Interesse an Phagen und DNA verloren, weil ja Watson und Crick den Kern des Pudels geknackt hatten?

Ich selbst blieb der T1-Genetik treu und hatte bald in meinem Vetter Thomas Trautner einen Mitarbeiter, der sich speziell den Heterozygoten widmete. Dann ging ich ein Jahr nach Brasilien, um in Rio ein Phagen-Labor einzurichten. Mein Hauptgewinn aus dieser Zeit war die Bekanntschaft mit dem jungen schweizer Mediziner Rudolf Hausmann, aus der eine lebenslange Freundschaft und Zusammenarbeit wurde.

Weit ab vom Schuß, in Südamerika, merkte ich kaum, daß Max drei entscheidende Sommer-Monate in Köln verbrachte, in denen schwerwiegende Pläne geschmiedet wurden. Erst später erfuhr ich, daß Prof. Joseph Straub mit vermutlich langfristiger kluger Strategie Max Delbrück ein Extraordinariat für Mikrobiologie angetragen hatte. Straub gehörte zu den wenigen Hochschullehrern, die ohne Rücksicht auf die eigene Statur Kollegen von höchst möglichem Standard als Fakultätskollegen suchen. (Das vor allem ist das Geheimnis der Exzellenz einer Fakultät oder Universität, von der heute so viel geredet wird.)

Max kam zu Gesprächen, die zum ersten Phagenkurs in Köln führten, an dem — äußerst wichtig für die Zukunft — auch Peter Starlinger teilnahm. Bei dieser Gelegenheit schmiedeten Straub und Delbrück auch die Pläne für ein großes Institut. Delbrück lieferte sein weltweites Renommee als Beitrag — Straub sein Verhandlungsgeschick und seine tiefe Kenntnis der Bürokratie der deutschen Wissenschaft.

1956–64

Durch einen Brief von Bonhoeffer erhielt ich auf der Rückreise aus Südamerika erste Andeutungen über die Kölner Pläne. Schnell folgten — wohl von Straub — eine Einladung und bald ein Lehrauftrag für Mikrobiologie. Als ich in Köln eintraf, hatte Starlinger seine Arbeit dort schon begonnen. Sofort entwickelte sich eine harmonische Zusammenarbeit in Freundschaft und gegenseitigem Vertrauen.

Nachdem auch Günther Stent — und ich glaube auch Weidel — das ursprünglich Max zugedachte Extraordinariat abgelehnt hatten, war die Reihe an mir. Zugleich erlaubte die Bürokratie auch Planstellen für Starlinger und Trautner. Rudolf Hausmann folgte aus Brasilien und zu aller Freude kam Gus Doermann aus Nashville für ein Jahr als Gastprofessor. Räumlich war unsere kleine Gruppe im Erdgeschoß des Botanischen Instituts untergebracht und in einem winzigen benachbarten Privathäuschen. Unsere Arbeit kreiste um Baubesprechungen, Unterricht und Labor.

Delbrücks Geist schwebte über den Wassern und dirigierte sanft aus der Ferne.

Max hatte ein weiteres Extraordinariat für Strahlenbiologie erkämpft, das Walter Harm 1958 annahm. Alles Übrige lief schleppend, der Baubeginn verschob sich ständig. Immer wieder tauchte in Max´ ungeduldigen Briefen der Satz auf: „Noch kein Wort aus Düsseldorf".

Sehr ungelegen kam dazu der Umzug von Straub als MPI Direktor nach Köln-Vogelsang. Sein Lehrstuhl wurde von Prof. Wilhelm Menke übernommen. Am 18.11.1957 schrieb Max: "I'm just about ready to write the whole thing off".

Meist aber war die Korrespondenz wissenschaftlich. Max war als Kritiker begehrt — sein Urteil war hart — oft schroff bis verletzend. "This thing will explode right into your face", schrieb er einmal. Oder, als ihm eine Schlußfolgerung nicht paßte: „Wenn ich nicht so ein starkes Vorurteil hätte, wäre ich von deinen Argumenten überzeugt — aber ich habe nun mal dieses Vorurteil". Schwer dagegen zu diskutieren. So dauerte es auch lange, bis Max eine Korrektur an seiner mating-theory akzeptierte. Dazu brauchte es die vereinten Kräfte von Frank Stahl, Charly Steinberg und uns.

Es kamen wichtige Gäste, um zu sehen, wie sich Delbrücks Institut entwickelte. Alle mußten Seminare geben, für die Peter unentbehrlich war. Mit der Begründung, abschnittsweise deutsche Übersetzungen für die Studenten zu geben, lieferte er prägnante Kurzfassungen, oft angereichert mit Zusatz-Informationen, die mir oft zu besserem Verständnis halfen. George Beadle besuchte uns für mehrere Vorträge und antwortete auf die Frage, ob er auch mit Pflanzen arbeite, nur: "Plants - ? Plants have too many molecules."

Als Leo Szilard das neue Delbrück-Institut besuchen wollte, stand das immer noch im Rohbau. Dieser faszinierende Mann, der einst Einstein veranlaßte, bei Roosevelt den Bau der Atombombe anzuregen, stellte Fragen am laufenden Band – ihn interessierte einfach alles: der deutsche Forschungsbetrieb, die Parteien-Situation, unsere Laborbedingungen, aber vor allem die politische Stimmung – während des Krieges und im Nachkriegsdeutschland. Dazu ohne Bezug schrieb er in mein Buch: "Biologists — so I'm told — fall into two classes: the 'sons-of-a-bitch' who write papers and put in things which are not so and the 'bastards', who point this out."

Unsere kleine Gruppe vergrößerte sich ständig — viele von damals sind heute hier. Ich kann sie nicht namentlich aufzählen und schon gar nicht zeitlich einordnen. Im Herbst 1958 ging Peter für ein Jahr nach Pasadena. Als er zurückkam, war der Bau immer noch nicht fertig, aber Peter rückte — es war wohl damals — offiziell in die Funktion eines Abteilungleiters auf, die er de facto schon seit langen innehatte.

Auch Max hatte dem Druck des Ministeriums nachgegeben und den Ruf nach Köln endgültig angenommen. Vielleicht beschleunigte das auch den Institutsbau. Denn als auch Thomas Trautner nach einem Post-doc-Jahr bei Arthur Kornberg zurückkam, war im Herbst 1961 endlich der Neubau bezugsfertig. Die offizielle Einweihungsfeier fand aber später, am 22.6.1962 statt. Max hatte zu diesem Anlaß seinen alten Lehrer und Freund Niels Bohr als Hauptredner zu einem Festkolloquium eingeladen. Der Vortrag war physikalisch-philosophisch-historisch. Ich erinnere mich nicht, wieviel ich davon verstanden habe — aber es war sehr erhebend.

Der Institutsbetrieb lief jetzt auf vollen Touren. So fand im März wieder ein Phagenkurs, bald auch ein Bakterienkurs statt. All diese Kurse liefen nach dem in Cold Spring Harbor (CSH) seit 1947 bewährten Rezept ab: Morgens um neun eine Tasse Kaffee und ein Seminar von einem auswärtigen Gast, einem Mitarbeiter des Hauses oder auch von einem der Kursteilnehmer. Anschließend dann Laborversuche und nachmittags Auswertung — meist vom Vortag.

Bemerkenswerterweise nahm Max jeden Morgen an den Seminaren des Kurses teil, zeigte aber sehr zum Ärger der Vortragenden sein Desinteresse durch penetrante Dauervertiefung in die Morgenzeitung. Als am letzten Kurstag Max selbst den Seminar-Vortrag halten wollte, hatte ein Spaßvogel (Fritz Melchers?) am Eingang einen Stapel Zeitungen deponiert. Jeder verstand. Max machte gute Miene zu bösem Spiel und redete unbeirrt bis eine Zeitung nach der andern heruntersank.[57]

Diese Kurse halfen sicher zur Verbreitung neuen Wissens im Land, waren zugleich aber auch für das Institut recht hilfreich, um geeigneten Nachwuchs kennenzulernen. Aber auch Kollegen aus dem Haus, wie der als Abteilungsleiter für Biochemie inzwischen gewonnene Hans-Georg Zachau nahmen teil. Auch Pamela Abel, unsere neuseeländische Virologin, hatte ich bei so einem der beiden Phagenkurse, die ich in CSH halten konnte — einen der Kurse übrigens zusammen mit Wolfgang Michalke und Marie-Louise Reusse, damals „Lieschen", heute Frau Lengeler — kennengelernt. Aber auch aus anderen Kontakten gewann das Institut Mitarbeiter — so kamen 1963 Rainer Hertel und Karl Müller durch Seymour Benzer zu uns.

Höhepunkt des Jahres 1962 war der Besuch von Jim Watson, der nach Verleihung des Nobelpreises auf Europa-Tournee nach vielen anderen Vortragsorten völlig erschöpft und total heiser auch Köln als eine seiner letzten Stationen anlief. Bei seinem Vortrag lag er im Lesesaal des 6. Stocks ausgestreckt auf einem Tisch und flüsterte Peter Starlinger ins Ohr, der diesmal auch noch die Funktion des Lautsprechers übernehmen mußte.[58]

---

[57]Siehe auch den Beitrag von Fritz Melchers in diesem Band (d. Hg.).

[58]Siehe auch den Beitrag von Fritz Melchers in diesem Band (d. Hg.).

Max provozierte die Fakultät — und nicht nur diese — in Sitzungen gern durch gewollt unschickliches Benehmen. Auch 1962, in der *Spiegel*-Affaire, schockierte das Institut durch einen öffentlichen Aufruf zur Wahrung der Pressefreiheit, den mit Delbrück 28 Professoren und Dozenten aus Naturwissenschaft und Medizin unterschrieben hatten.[59]

In diesen Blütejahren 1962/63 verfügte das Institut über ein Ordinariat (Delbrück), zwei Extraordinate (Harm und Bresch) und zwei Abteilungsleiterstellen (Starlinger und Zachau) mit vielen sehr tüchtigen jungen Leuten in allen Abteilungen. Aber damit war der erste Zenith erreicht.

Die schwermütige Abschiedsphase begann mit der Abreise der Familie Delbrück selbst, die inzwischen auf vier Kinder angewachsen war. Trotz intensiver Bemühungen von vielen Seiten war Max nicht zu bewegen, auf sein geliebtes Caltech zu verzichten, das seinerseits keinesfalls mehr als zwei Jahre Urlaub zuließ. Während Max noch versuchte, Ulf Henning als neuen Abteilungsleiter zu gewinnen, wurden die Institutsmitglieder von außen offenbar als interessante Konkursmasse gesehen. Für alle von uns kamen lockende Angebote. Wir Betroffenen erwarteten eigentlich — als schnelle Gegenaktion der Fakultät — die Aufwertung unserer Positionen, um den personellen Ausverkauf zu stoppen. Ich habe nie erfahren, wer oder was die Fakultät damals daran hinderte. Erst als Walter Harm als full professor nach Baltimore ging und mich ein Ordinariat nach Freiburg zog, rang sich die Kölner Fakultät zu entsprechenden Gegenangeboten durch. Zu spät — es hatte sich viel Ärger angestaut.

Gleichzeitig gingen starke Rufe an Starlinger, Trautner und Zachau. Pamela Abel ging nach Australien und auch Ulf Henning wurde schon wieder aus Köln weggelockt. Das Haus leerte sich zusehens. Hier muß ich den roten Faden an Peter Starlinger geben, denn er war der einzige, der zunächst mit Rainer Hertel dieser Krise widerstand. Peter Starlinger allein war die Säule, die einen Wiederaufbau des Instituts möglich machte.

---

[59]Als Reaktion auf einen kritischen *Spiegel*-Leitartikel über die Bundeswehr, den Bundeskanzler Adenauer als einen Abgrund von Staatsverrat bezeichnete, wurden die Redakteure Conrad Ahlers und Rudolf Augstein verhaftet. Im weiteren Verlauf mußte Verteidigungsminister Franz-Josef Strauß zurücktreten.

4

The New Start

Peter Starlinger

Let us see what happened to the institute after Max Delbrück left. His departure was, of course, a great loss, but it was an anticipated loss. We hoped, however, that by this time the institute would be self-sustaining. After all, there was Carsten Bresch, the man who had built up the Institute's core since 1956 and who was next to Delbrück during the years of the latter. He was also the author of the textbook from which generations of German students learned their genetics. We all thought that he would become the successor of Delbrück, but we have just heard that this didn't work out: he left Cologne. The third professor at the Institute, Walter Harm, a pioneer in DNA repair, who had worked in those early years on photoreactivation and excision repair, had already left a little earlier and thus the three chairs that Joseph Straub had so lovingly collected for the institute were all vacant.

Consequently, the other group leaders were invited to fill the void. The first of these was Hans Zachau, who at that time was at the zenith of his fame. He was among the few people who at that time sequenced RNA; still a far cry from the human genome project but they had finished the sequencing of two t-RNAs very shortly after Robert Holley did this first. This was very much acknowledged everywhere and it would have been very good if Zachau had remained in Cologne. But he had already indicated that he might consider this only if he were not offered a chair in Munich, because Munich was the city where he would love to work and to live. He got that offer and that was that.

The next invitation went to Ulf Henning. Ulf Henning, originally a graduate student of Feodor Lynen, had worked in Stanford with Charles Yanowsky and had done spectacular experiments on the question of the co-linearity of proteins and genes. This is now so much a matter of course that it is hardly mentioned, but in those days it was an interesting question whether this would be true. And he had done the experiment to show that genetic recombination not only separated genes from each other or recombined parts of genes, but even recombined two allelic codons, thus creating a third codon. The codons were becoming known at that time from the experiments initiated by Heinrich Matthaei and Marshall Nirenberg. Amino acids in proteins could be determined by selecting appropriate tryptic fragments and sequencing them. Henning used two missense mutants that carried different amino acids in a particular tryptic fragment. By recombination, he obtained wild-type recombinants and showed that they carried an amino acid different from those of both mutants. The codon for this amino acid consisted of one nucleotide from one and two from the other mutant sequences. Thus the wild type arose by recombination within a codon. While Henning was in Cologne, he worked on a different system that he was about to develop. When he was asked, whether he would stay as a professor in Cologne, he was also offered the chance to succeed Wolfhard Weidel at the Max Planck Institute in Tübingen.

By the way, as we are talking about history, Weidel, together with Adolf Butenandt, had been the founder of modern biochemical genetics by showing long ago, in 1940, that 3-hydroxykynurenin was an intermediate in the formation of the red eye pigments of insects and that a mutant, isolated by Alfred Kühn, was defective in pigment formation because 3-hydroxykynurenin was missing. Weidel later switched to phage research and worked with Delbrück in Pasadena. He became a member of the phage group and later a director at the Max Planck Institute of Biology in Tübingen. But he died before he reached 50-years of age and Ulf Henning was offered his position.

Henning was a typical laboratory scientist; he loved to do his experiments himself. He didn't even want to have a large number of graduate students around him, not to talk about heavy teaching loads or administration. It was really no wonder that he accepted this position in Tübingen. By the way, being the director of an institute that was really not very large was still not to his liking. He cut it into two and lured Peter Overath from Cologne to Tübingen to become his co-director. During his years in Cologne, Overath had performed the brilliant experiments in which he had probed the function of the *E. coli* membrane by altering its lipid composition with the help of mutants specifically designed for this purpose.

At this time, the institute was really empty. Of the former group leaders, only Starlinger had remained and outside talent was dearly needed.

The first invitation went to Hartmut Hoffmann-Berling. Hoffmann-Berling, however, had a nice group in Heidelberg and was now offered a directorship at the Max Planck Institute. Why should he move? He did not. The same was true a little later with Friedrich Bonhoeffer of the Max Planck Institute for Virus Research in Tübingen, that stronghold of early German molecular biology.

The search process dragged on and on for years and the whisper grew that the Genetics Institute might be considered a failure and written off, the laboratory possibly being closed. However, at long last in 1967 a first score was made when Walter Vielmetter accepted an offer from the university and became a professor in Cologne. Walter had also been at the Max Planck Institute for Virus Research in Tübingen and had done very interesting work on the chemical mutagenesis of DNA. An even bigger catch was landed a year later when Benno Müller-Hill accepted an offer. Müller-Hill had at that time been at Harvard. He had, together with Walter Gilbert, isolated the lac repressor — the first regulatory molecule that was so prominent in the Jacob-Monod hypothesis, which was the centre of attention of the whole molecular biology community. When Müller-Hill accepted the offer from Cologne, the Institute made a big step forward.

From then on things improved. The next to be appointed was Klaus Rajewsky. Rajewsky had come to Cologne together with Ulf Henning but had not left with him. He had built up an immunology group which by now was very well acknowledged internationally. He got offers from the outside, and that allowed him to be appointed in Cologne in spite of the German rule that people cannot be promoted at their home university. Finally, Walter Doerfler came to Cologne. He had worked with David Hogness at Stanford on bacteriophage lambda and had switched to adenovirus at Rockefeller University. He now joined the Cologne Institute. By this time, the original hope of Max Delbrück had been fulfilled: there were five independent research groups. It was even a little bit better than he could have imagined, because all of these five people had tenured full professorships and thus could plan to work in Cologne for a long time, which all of them did until their retirement or even beyond.

This was good, but for research not only people are needed — you also need money. The time when the Petri dish was the major research instrument and when the running costs of the laboratory could be halved by replacing the glass lids of the Petri dishes with the metal lids of Nivea tins (obtained for free from a cosmetic company, a great invention of Carsten Bresch!) was now definitely over. The state of North Rhine-Westphalia has to be praised for providing professorships, but the budgets coming with them were small and resembled more those of an American research university than of a Max Planck Institute with its generous budgets and lush fringe benefits.

Obtaining sufficient research grants was obviously an important problem. Fortunately, at that time the *Deutsche Forschungsgemeinschaft* (DFG) had established a new funding scheme called *Sonderforschungsbereiche* (SFB), which were renewable big block grants of three years' duration. SFBs were awarded not to individual principal investigators, but to a whole group of them. This group had to assemble around a common theme, submit a joint application and show that the single projects complemented each other. We submitted an application and called this would-be SFB "Molecular Biology of the Cell". We were lucky; we were admitted to this scheme.

By the way, if I say "we", these were not only the professors of the Institute but also an equal or even larger number of *wissenschaftliche Assistenten.* Even today, we read in the newspapers that these "Assistenten" are nothing but the errand-boys of slave-driving professors until they are released by their "Habilitation" shortly before their retirement. This, however, was definitely not the case in Cologne. Better still, the DFG gave us the money to hire people from the outside for limited periods of time, but with a decent salary and the option to obtain a research grant through the SFB. Thus, we had, back in the 1970s, what is today called the "Junior Professor", who is supposed to rescue the German universities.

There were two little snags to the system. One was the stipulation that the first evaluation of the individual applications had to be made by the colleagues of the SFB themselves. It can easily be imagined that the judging of the nth application by the other n-1 members was a little delicate — to say it cautiously. This was later abandoned, if not formally, at least quietly, and the evaluation committees of the DFG took this task upon themselves.

The other problem was the declaration by the DFG that they would provide the money for a limited period only, and that after that the SFBs would have to be institutionalised by the state government. This, of course, did not happen, and the DFG compensated for this by extending the lifetime of the SFBs, in our case up to 18 years, followed by two new SFBs at the Genetics Institute, which allowed the funding to continue.

I should perhaps at this time also mention that the administration of the *Sonderforschungsbereich* was very good, because we had Anne Weber. She single-handedly did this job and channelled all these big sums of money into many small pots, and particularly made sure that no money would be left by the 31st of December, at which date, according to the Golden Rule of German public accounting, it would have been lost because it would have been proof that the money was not needed. Thus, the SFB was one pillar of our funding. Later, there would be another one. But before I come to that, I have to digress for a moment to say a few words about the Max Planck Institut für Züchtungsforschung in Vogelsang.

This was an old institute that had wandered around in Germany due to the vagaries and uncertainties of the time after the war and had finally settled down in the outskirts of Cologne, some 7 km from here. At the time I'm talking about it was headed by Joseph Straub, of whom you have heard much, and Wilhelm Menke. Both of them had left a chair of botany at the University for the greener pastures of a Max Planck Institute. Now, time had come for them to retire. It was a big question whether the Max Planck Society would keep the institute alive. What they did was take a daring decision, wisely guided by Georg Melchers, who was at that time the doyen of Max Planck's botany and also of the committee that had to make the suggestions. They decided that the institute would not only be continued but it would be increased from two to four departments and it would be steered in the direction of the newly emerging plant gene technology.

The first director appointed was Jeff Schell; certainly not a breeder, not even a botanist, but a microbiologist from Gent, Belgium. In the 1970s Schell had found that a bacterium, *Agrobacterium tumefaciens,* could transfer some of its genes into the nucleus and even into the chromosomes of plants. He was now vigorously developing this observation into a technique for the general gene transfer into plants that would revolutionise plant breeding and agriculture.

The next to be appointed was Heinz Saedler. He was at the time a professor of genetics in Freiburg, and before that had spent many years in Cologne. He was also working with bacteria, but you will hear from him tomorrow that he could use this as a very sound basis for doing very innovative experiments in plant development and evolution. He was followed by Klaus Hahlbrock, a biochemist from Freiburg. And last came Francesco Salamini from Bergamo, Italy, a breeder with a strong interest in molecular biology. The scope of Cologne molecular biology had thus increased by at least a factor of two, which was of course very welcome.

However, there were more immediate benefits for the Institute of Genetics. Once upon a time — you notice that I'm going to tell you a fairy tale from the distant past — once upon a time, there was a German research minister who considered plant gene technology not as a liability but as an asset. This minister approached Schell and asked whether he would not like to form an institution that could be called a Gene Centre, thus improving and enlarging the Institute with federal money. Schell agreed immediately. He suggested, however, including the whole of the molecular biology community in Cologne, meaning the Institute of Genetics with its predominantly animal and bacteria-oriented research. The minister agreed. The structure was organised pretty much like a SFB, but in this case it was called a "Gene Centre". This Gene Centre became the second important pillar of our funding.

Thus, people were there and money was there, and the consequences were foreseeable: the Institute was expanding. Post-docs came from all over the world and so came many graduate students, most of them from Cologne but some also from other German universities. The latter were supported by a graduate programme initiated by the Fritz Thyssen Foundation and later on by the DFG. As a consequence, the Institute became very crowded. More and more incubators and ultra-centrifuges and other equipment appeared in the halls and in the staircases, and one day the fire department informed us that this would have to be remedied or the Institute would be closed down for being dangerous.

This was reported duly to the ministry and the answer we received was cool: they said that our space was certainly sufficient for the people provided by the university and by the state of North Rhine-Westphalia. If on top of that we created a big *Sonderforschungsbereich*, that was our private business and, if necessary, we would have to reduce it or to close it down. This was certainly not what we had hoped for. But who would help? Could anybody help? This time it was not the DFG, because it is a public institution. It has a charter and while it was always very helpful within this charter, there were limits. Building new facilities — an annex to the Institute — was beyond its limits. It was the *Stiftung Volkswagenwerk* that came to our rescue; it had a slightly more liberal charter. It was not easy for the *stiftung* either, but it was possible and it gave us money for an annex. Those of you who know the old institute building on Weyertal may have noticed that this grim, grey building has a red and white rear end. This red and white structure was the annex financed by the *Stiftung Volkswagenwerk.*

This helped a lot, but it was still not sufficient. However, we were lucky again. One day the directors of the Institute — I am telling another fairytale — the directors of the Institute were invited to meet the management of the Bayer company. We were well received on top of the high-rise building in Leverkusen and treated to an opulent meal. This in itself would not warrant being included in the history of the Institute, but the CEO of Bayer announced during that meal that they would give us a gift of one million Deutschmarks. Thus, if you have learned that there is nothing like a free lunch, this was an exception.

We were heavily criticised for that later and it was said we were selling out the university to the profit of industry; that we were in Bert Brecht's parlance, the *"erfinderischen Zwerge, die für alles gemietet werden können"* ("inventive dwarves that can be hired for everything"); but it was simply not true. We got the money without any strings attached, and this money was very welcome because part of it could be used for revamping and improving Klaus Rajewsky's mouse facilities, which was dearly, dearly needed, and the rest could be used for a greenhouse that the plant group had to have, not only for doing our experiments. It was also necessary because always there lurked the danger that during these experiments a monster could appear, which in German parlance is called a "gene plant". And a gene plant outside of a closed building was absolutely impossible. So part of the money could be used for this purpose.

There was, however, the problem of where to build the greenhouse. Every square foot of ground outside our building belonged either to Botany or to Zoology and obviously a Genetics greenhouse could not be placed there. It was again the Max Planck Institute that came to our help. We were allowed to build the greenhouse on the premises of the Max Planck Institute, and even more than that followed. Heinz Saedler, supported by his colleagues, offered the plant group in the university laboratory space at this institute. Several people who at that time belonged to the Institute for Genetics are still around, such as Wolfgang Werr, who is now a professor in developmental biology, George Coupland, who is the successor to the late Jeff Schell, and Klaus Theres who also heads a group at the Max Planck Institute. They all benefited from this offer and could do the experiments in Vogelsang.

During these years another joint venture between the University and Max Planck was started, the institution was called the Max Delbrück Laboratory in honour of our first director. It was a new laboratory building for six junior research groups that could work there for a limited period of time. There were two generations of this. The running money came from the Federal Ministry, and three of these groups worked in the fields of animal or yeast genetics. The leaders of these groups were nominated by the University. The other three groups worked on plant genetics and these were selected by the Max Planck Institute. At the end of this period that I am covering here, there was even a first section of the new Genetics Institute built, so that the groups of Wolfgang Wille and Boerries Kemper could move over there. Thus, the lab space situation that had been so problematic for a time was gradually not only defused but even considerably improved.

Most of this time, and for as long as he could, Max Delbrück took a lively and friendly interest in our Institute. He came to visit us, he talked to younger and sometimes also older members of the Institute, and he asked how they were doing and what they were doing. He also gave lectures, not only about his own *Phycomyces* research, but also on topics that interested him at the time and that he thought were interesting for us. I remember one such series on the molecular clock in evolution which was of obvious interest to geneticists and another such series on ferromagnetism and the Ising model, which sounded to us a little bit far-fetched, but which may well prove to have been far-sighted today, at a time of neuron networks and synaptic plasticity.

Max Delbrück continued these activities even when he fell seriously ill with leukaemia towards the end of the 1970s and until we had to mourn his death in 1981—the death of a mentor not only of our Institute, but of much of post-war biology in Germany. He acted in this role first in Berlin and in Göttingen, then in Cologne, and later in Konstanz, and now he was gone, very much missed by all of us.

There was another death to deplore. Ten years later Wolfgang Wille, a young professor at the Institute who, together with his group, worked on the genetics of the brain, died suddenly and completely unexpectedly. He suffered a stroke while giving a lecture to the students, fell into coma, and failed to wake up again. To see a young person who was so energetic and so enthusiastic pass away at the age of just over forty, where we all thought he would have 20 or 30 years in front of him to carry out his many plans, was very sad indeed.

Let me summarise. What I tried to show was that with the help and the support of the people and of the institutions mentioned above, and certainly also with some luck, the Institute was able to provide an environment in which its members could teach but also do research at a scope that allowed them to participate a little bit in that great and lasting achievement of the second half of the last century, the development of molecular biology. To have been around at that time was very enjoyable and very gratifying and I can only express my hope that for the people now working at that Institute, the first half of the 21st century will be as rewarding as the last one was for my generation.

5

Building Molecular Biology in Post-War Europe: Between the Atomic Age and the American Challenge

Bruno J. Strasser

Introduction

The creation of the Institute of Genetics at the University of Cologne was a unique event, and yet, it was typical of broader changes that took place in the life sciences in Europe around 1960. In order to situate the history of the Cologne Institute of Genetics in a larger context, I will outline similar initiatives in other European countries, and show how the development of molecular biology was related to broader cultural factors such as the Atomic Age and the American challenge.

The Institute of Genetics in Cologne was created in 1962.[60] The project was planned and organised by Joseph Straub, a botanist at the University of Cologne, and Max Delbrück, a phage geneticists working at Caltech. In 1956, Max Delbrück explained that he wished to accomplish something that "neither Lwoff [in France], nor Maaløe [in Denmark], nor Hayes [in the United Kingdom] have been able to accomplish, namely to carry modern biology right into a university

[60]Ernst Peter Fischer and Carol Lipson, *Thinking about Science. Max Delbrück and the Origins of Molecular Biology*, New York 1988.

set-up".[61] Max Delbrück added that "modern biology" had developed essentially in research institutions, such as the Max Planck Institut für Virusforschung in Tübingen, the MRC [Medical Research Council] laboratories in the United Kingdom or the Pasteur Institute in Paris. It was indeed rare to find these new approaches to biological research being carried out in European universities.

Meanings of molecular biology

What did "modern biology", or "molecular biology" as it became increasingly called, mean in the late 1950s? The term "molecular biology" was used, probably for the first time, in 1938, in an annual report of the Rockefeller Foundation. It designated a large funding programme for different kinds of research in the life sciences which relied on a combination of physical and chemical methods.[62] This definition was extremely broad and only overlapped partially with the much narrower meaning that "molecular biology" took in the 1950s.

At that time, molecular biology did not exist as a unified discipline with clear-cut boundaries. For Max Delbrück and many others at the time, "modern biology" was understood as a new way of doing research, of asking questions and combining methods. It represented, for example, a new way of using heavy instrumentation in biology, such as ultracentrifuges, electrophoresis apparatus and electron microscopes. It also involved working with different kinds of organisms and new experimental systems, such as phage, bacteria or *Neurospora*. New concepts — "jargon," said the critics — such as "information" or "genetic programme", were distinctive of the new biology. Research in "modern biology" often relied on interdisciplinary collaboration between geneticists, biochemists, microbiologists or biophysicists, for example, something that was rather unusual in the life sciences at that

[61] Max Delbrück to George Beadle, 6 September 1956, Delbrück papers, folder 2.10, Caltech Archives.
[62] Rockefeller Foundation. *Annual Report.* New York 1938.

time. According to Max Delbrück, the institutional structures of the universities prevented the development of interdisciplinary research. Indeed, creating an institute of modern biology in a university meant "breaking down the organisations deadlock in which biology finds itself all over the world, with Caltech as an almost unique exception". This deadlock resulted from "the fact that departments of botany and zoology were created long ago" and had prevented "the growth of genetics, biochemistry, microbiology, physiology, etc. in the academic institutions".[63]

Only around 1960 did these different approaches coalesce into a unique disciplinary framework, under the name of "molecular biology". Research focussed on the structure, the function and the relationships between two classes of molecules, proteins and nucleic acids. A number of institutions were created to accommodate this new approach to biology. This transformation also had broader cultural meanings that can help us understand why and how molecular biology came to be understood as "*modern* biology" around 1960. By identifying these larger contexts, I hope to shed some light on the creation of the Institute of Genetics in Cologne.

Institutional developments in Europe

The creation of an Institute of Genetics in Cologne was typical of a much broader institutional trend. The first laboratory in the world to carry the name "molecular biology" was the MRC Unit for Molecular Biology in Cambridge, a new name given in 1957 to the MRC Unit for the Study of Molecular Structure of Biological Systems. A year later Francis Crick and Max Perutz, two crystallographers, were making a proposal for a large institute in Cambridge which eventually became, in 1962, the Laboratory of Molecular Biology. What did "molecular biology" mean in Cambridge? A combination of crystallography and biochemistry, bringing the work of Francis Crick, Max

---

[63]Delbrück to Beadle, September 6, 1956, Delbrück papers, folder 2.10, Caltech Archives.

Perutz, John Kendrew, Fred Sanger and Sydney Brenner under the same roof.[64]

In France in 1960, different initiatives were taken by a new science policy agency of the Fifth Republic, the DGRST [Direction Générale de la Recherche Scientifique et Technique], to support promising fields in science. Molecular biology was chosen as one of them. It was understood as a combination of microbiology, genetics and biochemistry, and the persons who were most active in shaping this understanding were Jacques Monod from the Pasteur Institute and Raymond Latrajet. Thus, in Paris, molecular biology had a rather different meaning than in Cambridge.[65]

In Switzerland, the first institute for molecular biology was created in 1962 around very different research traditions: biophysics (mainly electron microscopy and phage genetics) and biochemistry. It was headed by Eduard Kellenberger, a pioneer in electron microscopy and phage genetics, and Alfred Tissière, a biochemist.[66]

Finally, the Institute of Genetics in Cologne was created with a strong emphasis on genetics and biophysics. These different cases — and one could add other examples from Naples and Brussels — highlight that the creation of institutions devoted to molecular biology was a general trend around in Europe 1960, even if "molecular biology" assumed quite different meanings in each case. One might wonder why this institutionalisation process took place specifically around 1960.

Part of the answer might be related to money. The funds available for research from the different national funding agencies in Europe in the four countries I have just examined increased dramatically during the late 1950s.[67] This increase came in part as a result of the

---

[64]Soraya de Chadarevian, *Designs for Life. Molecular Biology after World War II* (Cambridge 2002).

[65]Jean-Paul Gaudillière, *Inventer la biomédecine : la France, l'Amérique et la production des savoirs du vivant : 1945–1965* (Paris 2002).

[66]Bruno J Strasser, La fabrique d'une nouvelle science: La biologie moléculaire à l'âge atomique (1945–1964) (Florence 2006).

[67]Bruno J. Strasser, "Institutionalizing molecular biology in post-war Europe: a comparative study", *Studies in the History and Philosophy of Biological and Biomedical Sciences* **33C** (2002), pp. 533–564.

post-war economic reconstruction. However, if European governments did indeed have more money to spend, this does not explain why it was spent on molecular biology, rather than on other fields of science. Indeed, the priorities of European scientific policy bodies in the mid-1950s were mainly atomic physics and space research, not molecular biology. The reason why molecular biology was so successful in that policy environment was in part due to its links with physics, and biophysics in particular. In the context of the Atomic Age, physics enjoyed an extraordinary prestige, scientifically and culturally.

Atoms for peace

Even though atomic energy symbolised the threat of nuclear war, it also seemed to hold innumerable promises of technological, scientific and social progress. In 1946, a French equivalent of *Scientific American* was published under the name *Atomes, tous les aspects scientifiques d'un nouvel age*, indicating the privileged place of atomic physics in the post-war era. Only at the dawn of the Atomic Age, in 1970, was the title changed to its present name *La Recherche*. Similarly, the main symbol of the 1958 Universal Exhibition in Brussels was the Atomium. Apart from the restaurant, all the spheres of this astonishing construction were devoted to exhibitions on the peaceful uses of atomic energy.

Only five years after Eisenhower's 1953 Atoms for Peace initiative, funding bodies were giving a strong emphasis to research related — even remotely — to the peaceful uses of atomic energy. All kinds of research at the frontiers of physics and the life sciences, in genetics, radiobiology, nuclear medicine or biophysics, benefited immensely from this particular context. Understandably then, the promoters of molecular biology often emphasised the association between their new science and biophysics. In Cologne, this link was made all the more obvious since the project was promoted by Max Delbrück, trained as physicist whose research was considered to fall under the heading of biophysics. The association between molecular biology and biophysics provided a powerful political and cultural justification for new discipline during the Atomic Age.

The American challenge

The second point I would like to make is that about the relationship of molecular biology and American science. In post-war Europe molecular biology was often portrayed and perceived as an American science and the cultural value of America, like that of physics, was extremely high in the 1950s. The United States were often considered as a model for industrial and scientific management. The rise of the consumers' society, based on American products, argued forcefully that the United States had found a unique key to bolster its scientific and industrial productivity.

"America" was not only a symbolic and rhetorical resource, but also a material and intellectual one. Transatlantic exchanges between Europe and the United States played a crucial role in all places where molecular or modern biology developed in post-war Europe. This geography of exchanges was unusual. In the inter-war period, it was common for a young scientist to spend a post-doc in Germany, but in the post-war period it became increasingly important to go to the United States. This transformation did not happen all at once. It reflected a progressive change in the understanding of where the avant-garde of scientific progress was. Molecular biology grew essentially in a trans-national space where American and European researchers could meet, collaborate, and exchange ideas.[68] In the immediate post-war period a number of European scientists were extremely isolated from the United States, for material, cultural or

[68]Pnina Abir-Am, "From multidisciplinary collaboration to transnational objectivity: International space as constitutive of molecular biology, 1930–1970", in E. Crawford, T. Shinn and S. Sörlin (eds.), *Denationalizing Science – the Contexts of International Scientific Practice* (Dordrecht 1993), pp. 153–186.

political reasons, as Ute Deichmann has shown in the case of Germany.[69]

European refugees who had left Europe for the United States in the 1930s played a central role in overcoming this isolation when they eventually returned to the country of origin. Max Delbrück for Germany, Severo Ochoa for Spain or Jean Weigle for Switzerland all provided crucial links to the United States for European laboratories in the post-war period.

America was a central reference point for science policy bodies in Europe in the post-war period. The Nobel Prize attributions did in fact make it very evident that Europe as a whole had lost its pre-eminence in science. In the 15 years preceding the war (1925–39), France, Germany and Britain had shared 32 Nobel Prizes in the natural sciences (physiology or medicine, chemistry, physics), whereas the United States had only received 11. But in the 15 years following the war (1946–60), the United States had largely taken the lead, with 38 Nobel Prizes, compared with only 18 for the three European countries. It was obvious for anyone who looked at these figures that the weight of scientific research had shifted across the Atlantic.

At the same time scientists and science policy officials promoted the idea that there was an "American challenge" to meet: "*Le défi américain*" for the French, or what came to be known as the "scientific and technological gap". In science, Europe suffered, supposedly, from a "historical backwardness", a delay compared to the "normal" historical development of the life sciences. This strategy made it possible to advocate the necessity for "catching up". In this context, molecular biology was supported by science policy bodies in the name of modernisation of science in Europe and the catching up with the United States.

---

[69]Ute Deichmann, "Emigration, isolation and the slow start of molecular biology in Germany", *Studies in the History and Philosophy of Biological and Biomedical Sciences* **33C** *(3)* (2002), pp. 449–471.

Conclusions

These two contexts, the Atomic Age and American science, help us understand why molecular biology became established as a new science in Europe around 1960. The creation of the Institute of Genetics in Cologne was unique yet, in another way, it reflected a broader transformation of science and society in post-war Europe. The opening of the institute in Cologne took place approximately at the same time as other institutes devoted to molecular biology elsewhere in Europe. Therefore, the case of the Institute of Genetics in Cologne should not be understood as a story of catching up with institutional developments that had taken place much earlier in other European countries.

Local historical studies are necessary to identify these larger trends and understand how they played out at the local level. Max Delbrück explained in 1956: "What appealed to me so strongly to the Cologne proposal is partly the attraction of the vacuum".[70] Unlike what Delbrück suggests, I tried to argue that new scientific traditions never grow in a vacuum, but always within specific cultural and political contexts. This is why the development of science can only be understood by situating this process within the broader changes that take place in society.

[70]Max Delbrück to George Beadle, September 6, 1956, Delbrück papers, folder 2.10, Caltech Archives.

6

Beitrag zur Gedenkfeier für Max Delbrück, 10. März 1982[71]

Joseph Straub

Herr Starlinger bat mich, ich möchte heute etwas von den persönlichen Begegnungen mit Max Delbrück, wie sie zur Errichtung des Instituts für Genetik führten, also aus der Frühzeit der Kölner Genetik, erzählen. Ich tue dies gerne.

Zuerst möchte ich das Bild von Max Delbrück, das sie soeben aus verschiedenen Blickwinkeln gezeichnet sahen, in einen Rahmen einfügen. Nach allen Erlebnissen scheint mir ein von Herzen kommendes „Dankeschön" die weitaus passendste Umrandung zu sein: Dank für die selbstlose Hilfe und Unterstützung, die er der Kölner Biologie, damit unserer Universität und wohl auch der Biologie in Deutschland, durch das Zustandebringen dieses — wie ich als jetzt abseits Stehender sagen darf — dieses wirkungsvollen Instituts geliehen hat; Dank aber vor allem für alle Freundlichkeit, darunter viel Frohsinn, und die Freundschaft, die er uns schenkte. Dank auch für eine beeindruckende Wahrhaftigkeit; sie machte ihn zu einem Menschen, auf dessen gegebenes Wort man sich fest verlassen konnte.

Ich erfuhr den Namen Max Delbrück zum Mal bald nach meiner Dissertation. Mein Lehrer in Freiburg, Friedrich Oehlkers, und Hans

[71]Lecture by Joseph Straub on the occasion of the Max Delbrück memorial at the Institute of Genetics in Cologne on 10 March 1982, Archive of the Institute of Genetics. Emphases in the original.

Spemann, der Freiburger Zoologe, Entwicklungsphysiologie am Amphibienkern betreibend, Nobelpreisträger, veranstalteten im WS 37/38 ein gemeinsames Kolloquium. Ich bekam damals von Oehlkers die Aufgabe zu einem Referat über eine Publikation von Max Delbrück, Nikolay Timofeeff-Ressovsky und Karl Zimmer mit dem Titel „Über die Natur der Genmutation und der Genstruktur". Dieses abendliche Kolloquium, in einem kleinen Hörsaal, wo Spemann neben Oehlkers, wie immer, in der vordersten Reihe saß, ist mir aus einem banalen Grund im Gedächtnis geblieben: Spemann schlief während meines Referates ein. Tags darauf äußerte Oehlkers, es sei ihm unverständlich, wie man bei einem solch wichtigen Thema, welches die grundlegenden Probleme der Biologie beträfe, einschlafen könne. Aber Oehlkers vergaß, daß Spemann Frühaufsteher und meine Darstellung schlecht war. Ich konnte diese Veröffentlichung nämlich nicht ganz verstehen, da sie viele physikalische Überlegungen und mir unbegreifliche Formeln enthält. Aber der Versuch, etwas über das Gen auszusagen, blieb in meinem Kopfe mit dem Namen Delbrück verbunden.

In den Jahren zwischen 45 und 50 erfuhr ich dann durch [Georg] Melchers von den Phagenarbeiten Max Delbrücks, die damals einen genetisch interessierten Biologen geradezu in Entzücken versetzen konnten. Max hatte darüber bei Besuchen in Tübingen vorgetragen. Hinzu kam, daß wir in den Jahren um 1950 bis 52 in zunehmendem Maße Kenntnis von den großartigen Forschungsergebnissen erhielten, die vor allem in den U.S.A. in der Pilz-, Bakterien- und Virusgenetik erzielt worden waren. Jedem Einsichtigen war es bald klar, daß dieses neu entwickelte Wissenschaftsgebiet in jeder Universitätsbiologie heimisch werden müßte, wenn sie nicht hoffnungslos nachhinken möchte. Aber wie könnte diese Einsicht in die Tat umgesetzt werden? Damals gab es in Köln drei biologische Lehrstühle, nämlich je einen für Zoologie, Entwicklungsphysiologie und Botanik. Da kam uns zu Hilfe — Sie hören richtig — das Bayrische Kultusministerium! Es berief den damaligen Ordinarius für Botanik von Köln [Straub] nach München und die Münchner hatten Glück, denn der Berufene blieb in Köln. Dafür richtete das Nordrhein-Westf. Kultusministerium einen Lehrstuhl für Mikrobiologie ein, der dem Botanischen Institut zugeordnet werden sollte. Am 15.02.55 beschloß die Philosophische Fakultät, der damals noch die Geistes- und die Naturwissenschaftler angehörten, auf meinen Vorschlag hin in einer ihrer letzten Sitzungen die Berufungsliste für den Lehrstuhl für Mikrobiologie, genetischer Arbeitsrichtung:

1. Max Delbrück, Pasadena
2. Wolfhard Weidel, Tübingen
3. Gunther Stent, Berkeley

Auf seine Berufung antwortete Max Delbrück postwendend — wenn ich vom Kanzler der Universität richtig informiert wurde — mittels einer Postkarte. Er bedankte sich und bedauerte, dem Ruf nicht Folge leisten zu können. Gleichzeitig teilte er aber mit, daß er sich auf Wunsch zur Verfügung stelle, um während eines Gastsemesters einen Kurs über Bakterien- und Phagengenetik abzuhalten und diesen mit Vorlesungen zu begleiten. Von da an war für mich die Fortsetzung jener Berufungsliste zweitrangig. (Später wurde der Lehrstuhl auf meinen Wunsch dem Institut für Genetik zugeordnet). Wir taten alles zum Gelingen des Gastseminars, und im Sommer 56 war es dann soweit. Linskens, mein tatkräftiger Mitarbeiter, mit seinem kleinen Fiat 600 und ich selbst mit dem VW-Käfer konnten an einem Apriltag jenes Jahres 56 Max mit Manny und den Kindern Jonathan und Nicola auf dem Düsseldorfer Flugplatz willkommen heißen. Sie waren guter Dinge, etwas müde zwar, aber fröhlich. In Köln-Frankenforst, wo wir eine Wohnung im Hause einer Fabrikantenwitwe gemietet hatten, wurden die Koffer abgestellt und die Eltern Delbrück gingen einkaufen. Ich erhielt von Max die Aufgabe, Jonathan und Nicola, damals wohl 8 bzw. 6 Jahre alt, solange zu beaufsichtigen. Das sollte sich als ein schwieriger Auftrag erweisen. Wir waren nämlich der Einladung der Hausbesitzerin gefolgt und hielten uns in einem sehr geräumigen Zimmer mit einer sehr großen Anzahl von erstklassigen Polstersesseln auf, die in Gruppen um Tische gereiht waren. Da wurden meine beiden Delbrücks munter und begannen auf den Sesseln Nachlaufen zu spielen. Die gute Polsterung verhalf zunehmend zu Übungen in der

Vertikalen. Es entwickelte sich schließlich eine wilde Hüpfjagd, während der die beiden natürlich Freudenschreie ausstießen. Die Vermieterin, eine betagte Dame, erschien, fiel fast in Ohnmacht und flehte mich händeringend um Abstellung des für sie grausamen Spiels an. So etwas lag mir aber nicht, zumal sich bereits auch die normale Ermüdung bemerkbar machte. Endlich kamen die Eltern zurück und ich unterrichtete Max über den Vorfall. Da vollbrachte er eine diplomatische Meisterleistung. Er ging mit Jonathan und Nicola in die Nähe der erbosten Hausbesitzerin und erklärte den beiden mit ernstem Gesicht in ausgesuchtem Deutsch etwas dieses: „Ihr müßt nun stets beherzigen, dass Ihr in einem Land eingetroffen seid, das auf der ganzen Welt vor allem wegen seiner überwältigenden Disziplin berühmt ist; in erster Linie müßt Ihr das heilige Gesetz befolgen, Polstersessel nur zum Sitzen in sauberen Kleidern zu benutzen usw." Jonathan und Nicola blickten an ihrem Vater gläubig aber sprachlos empor, sie verstanden kein Wort. Die Vermieterin jedoch war überzeugt, die besten Mieter gewonnen zu haben.

Jonathan und Nicola mußten in die Volkschule. Damals existierten nur konfessionell ausgerichtet Schulen. Max wollte die beiden deshalb bei der protestantischen Volksschule melden, stellte dem betreffenden Rektor aber zunächst die Frage, ob die Kinder in dieser Schule beten müßten, in dem Sinne, daß sie unter Strafandrohung dazu aufgefordert seien. Der Rektor bejahte dies entschieden, worauf Max nicht anmeldete. Wir besprachen die eingetretene Situation und schlugen vor, es einfach bei der Konkurrenz zu versuchen. Hier hatte Max Glück. Er traf nämlich auf einen echten Kölner als Rektor, also auf einen jener glücklichen Menschen, die dem lieben Herrgott mehr zu tun überlassen als sich selbst. Auf die wieder gestellte Frage nach dem Gebetszwang erklärte dieser Rektor nämlich: „Nä, nä, wer nicht beten will, soll's halt bleiben lassen." So gingen die Kinder in die Konkurrenzschule. Nun muß man wissen, daß sich all' dies zu Beginn des Monats Mai abspielte, dem Monat, der in der katholischen Kirche speziell der Marienverehrung gewidmet ist. Einige Wochen nach der Anmeldung kam Max lachend – wer kannte nicht dieses charakteristische, zunächst verhaltene Lächeln, das sich plötz — lich voll öffnete — ins Institut und erklärte, gestern Abend hätten

Manny und er merkwürdig gefühlvolle Melodien aus dem Schlafzimmer der Kinder vernommen. Beim genaueren Hinhören hätten sie festgestellt, daß die beiden in ihren Betten mit großer Inbrunst Marienlieder gesungen hätten! So hatte Max durch sein beharrliches Bemühen um die Freiheit in der geistig-religiösen Kindererziehung schon früh Ökumene an der Basis ausgelöst, mit hörbarem Erfolg!

Aber nicht alles verlief so lustig. Bevor wir uns an jenem ersten Tag in Köln-Frankenforst verabschiedeten, saß man bei Saft noch etwas plaudernd zusammen. Da fragte Max in ziemlich schroffem Ton: „Wann kann ich mit dem Phagenkurs beginnen?" Die Frage war mir unangenehm! Damals war gerade der Neubau des Gebäudekomplexes an der Gyrhofstraße im Rohbau fertig, und im Erdgeschoß, wo der Kurs stattfinden sollte, waren auch die Fenster und Türen schon angebracht. Aber sonst war alles noch roh! Ich hatte mir anhand anderer Beispiele vorgestellt, daß sich ein Gastprofessor zunächst einige Tage in seiner Wohnung umsieht, dann in der näheren Umgebung und noch einige Tage in der weiteren Umgebung. Hinsichtlich Max Delbrücks ein schwerer Irrtum! Nun hatten wir zwar zahllose Pipetten für den Kurs erstanden, aber die vielen Wasserbäder, die Max in einem Brief u.a. als wichtiges Instrumentarium angefordert hatte, waren in der Werkstatt erst in statu nascendi. So beantwortete ich die Frage ausweichend und Verzögerungen begründend. Max unterbrach mich mit „Übermorgen fangen wir an". In Tag- und Nachtarbeit wurden alsdann die notwendigen Vorbereitungen noch getroffen und schließlich begann der Kurs pünktlich, übermorgen. Die Teilnehmer, nämlich fortgeschrittene Studierende des Botanischen Institutes, wissen noch heute, wie ich soeben feststellte, viel von diesem Phagenkurs zu erzählen. Max führte den Unterricht vorbildlich streng durch. Ich durfte auch teilnehmen. Damals war ich Dekan der Mathematischen-Naturwissenschaftlichen Fakultät. Im ersten Praktikum erhob ich mich gegen 11.00 Uhr von meinem Arbeitsplatz und sagte: „Herr Delbrück – damals siezten wir uns noch – ich wollte mal kurz hinüber zur Universität, um im Dekanat Unterschriften zu leisten. Seine Antwort: „Sie bleiben hier." Gerühmt wird noch heute, wie intensiv Delbrück jedes einzelne Arbeitspaar gesondert unterwies.

Jenes Praktikum veranstaltete natürlich auch einige Festchen, auf denen der Tanz durch Gesellschaftsspiele, die Max vorbereitet hatte, unterbrochen wurde. Zum Beispiel wurden kleinere Briefchen ausgeteilt; paarweise mußte man sich dann nach den verschlüsselten Angaben der Briefchen auf recht eigentümlichen, verschlungenen Wegen zu einem Ziel durchfinden, wo schließlich z.B. eine Bierflasche als Lohn gefunden wurde. An dem freundschaftlichen Verhältnis, das Schüler und Lehrer schließlich verband, konnte auch die strenge Abschlußprüfung nichts ändern, für die sich alle anhand ihrer Versuchsprotokolle vorbereiteten und dann z.B. gefragt wurden, warum der Himmel blau sei!

Jahre später versammelten sich die Teilnehmer des Kurses draußen in Vogelsang, als Max Delbrück uns auf der Rückreise von der Verleihung des Nobelpreises besuchte. An jenem Abend war Max geradezu ausgelassen, zumal sich auch heilige Mächte belohnend, aber auch mit der Rute eingefunden hatten. Max schrieb mir nachher aus Pasadena, dieses Treffen mit den Ehemaligen, die ich gerne die alten Kämpfer nenne, habe er besonders nett empfunden.

In jenem denkwürdigen Sommer 1956 konnte ich mit Max des öfteren Gespräche zu zweit führen. Ihr Inhalt bezog sich unter anderem auf die gerade damals bekannt werdenden genetischen Fortschritte, die nun wirklich jeden Biologen, ganz gleich was er von Genetik hielt, beeindrucken mußten. Zum anderen legte Max mir nicht nur einmal dar, welche Veränderungen im Bereich der Universitäts-Biologie ihm notwendig und sinnvoll erscheinen würden, um in unseren Universitäten die Genetik auf das anderswo erreichte Niveau zu heben, man darf sagen wieder zu heben! Das ließ den Gedanken reifen, an der Kölner Universität die Biologie durch ein Institut für Genetik, und zwar ein starkes Institut zu ergänzen. Die Verwirklichung dieses Planes in jenen Jahren, als die Universität gerade erst mit den Plänen zum Bau der Physikalischen Institute begann, als auch die anderen großen Fächer, z.B. die Chemie noch immer behelfsmäßig untergebracht waren, und neue Lehrstühle auch nicht ohne weiteres eingerichtet werden konnten, erschien geradezu unmöglich. Aber manchmal hat der Mensch eben Glück!

Tatsächlich standen die Sterne der Universität damals gerade günstig: Beim Rektor, Prof. Kauffmann, Kunsthistoriker, besaß originelle Forschung einen ganz hohen Stellenwert, so hoch, daß manche Kollegen es ihm übel nahmen (!), und der Kanzler, Friedrich Schneider, Jurist von Haus aus, war auch der Meinung, man müsse die Finanzierung der Universität differenziert durchführen und nicht einfach das Hergebrachte fördern. Bei Schneider, der uns im Mai des letzten Jahres leider ebenfalls für immer verlassen hat, kam hinzu, daß er sich in einzelnen Fächern, in der Biologie im besonderen, durch Lesen und Studium von geeigneten, also verständlich geschriebenen Büchern ein eigenes Bild von den Fortschritten der Wissenschaft zu machen bestrebt war. Ich weiß z.B., daß er die erste Auflage des kleinen Buches von Weidel „Virus, die Geschichte vom geborgten Leben", gelesen hat und davon begeistert war.

Ihm trug ich etwa Juni 56 vor, warum man nach meinem Erachten in der Kölner Biologie einen deutlichen und starken Akzent durch Errichtung eines selbständigen Instituts für Genetik setzen sollte, und frug, ob er dafür Chancen sehe. Schneider war diesem Plan von vorneherein zugetan, meinte aber: „Lieber Herr Straub" – wenn Schneider „lieber Herr Straub" sagte, wußte ich immer, daß er etwas wichtiges folgen ließ - „so etwas kann höchstens gelingen, wenn ein auf diesem Gebiet ganz besonders ausgewiesener Wissenschaftler den Aufbau und den Inhalt des Instituts plant und die Leitung übernimmt". Dabei dachten wir beide natürlich an Max Delbrück. Und weiter: „Für die Finanzierung kommt in der derzeitigen Aufbauphase der naturwissenschaftlichen Institute wohl nur die Deutsche Forschungsgemeinschaft in Frage". Ich besprach dies alles mit Max. Dabei war es entscheidend - dieses „entscheidend" gilt ohne jede Einschränkung - daß er unseren Wunsch, er möge sich für die vorgesehene Aufgabe zur Verfügung stellen, zu erfüllen versprach. Im Kontakt mit Herrn Schneider bereitete ich alsdann einen Besuch bei Prof. Kauffmann als dem Rektor der Universität vor. Max meinte, da müsse er wohl eine Krawatte anziehen! Ich erwiderte, schaden könne das wohl nicht! Als wir uns am Universitätseingang zu diesem Besuch trafen, vergaß ich nicht, die Krawatte zu loben. Darauf Max: „95 Pfennig, Kaufhof".

Es wurde ein langer Besuch. Der Rektor bat nach einigen freundlichen Worten Herrn Delbrück, ihm einen Einblick in das Forschungsgebiet zu vermitteln. Max saß bis dahin aufrecht in seinem Sessel. Nach der Bitte des Rektors legte er sich sozusagen lang, verschränkte die Hände hinter dem Kopf und blickte unverwandt an die hohe Decke des Rektorat-Zimmers. Ich saß im Hintergrund und konnte von da aus gut beobachten. Ich sah, wie der Rektor und der Universitätskanzler während dieser ziemliche Zeit dauernden Denkpause abwechselnd den stillen Denker und sich gegenseitig stumm und erstaunt anblickten. Aber dann gab Max ein ausgezeichnetes Symposium von bereits verstandenen und noch unverstandenen genetischen Grundlagen, die an Bakterien und Viren soeben erarbeitet waren, bzw. noch der Aufklärung bedurften. Dieses Privatissimum machte auf die drei Anwesenden starken Eindruck.

Dasselbe war der Fall bei einem Mittagessen, zu dem der Präsident der Deutschen Forschungsgemeinschaft, Professor Gerhard Hess, eingeladen hatte, wobei Max Herrn Hess und dem Generalsekretär der DFG, Herrn Dr. Zierold, im ausführlichen Tischgespräch darlegte, was man seines Ermessens tun müsse, um die molekulare Genetik in unserem Staat effektiv werden zu lassen. Dieses Treffen, das für das weitere Vorgehen in der Frage der Instituts-Errichtung sehr wichtig war, fand in Bad Godesberg im Gasthaus „Zur Lindenwirtin", auch „Ännchen" genannt, statt, wohin ich Max und Kanzler Schneider in meinem VW-Käfer — das waren noch Zeiten, als der Universitätskanzler im VW-Käfer fuhr — chauffiert hatte. An einem der folgenden Tage richtete ich an die DFG einen Antrag. Er enthielt u.a. die Finanzierung eines Instituts für Genetik nach den Vorstellungen von Max Delbrück, und ich durfte dessen Bereitschaft erklären, zum mindesten in den Anfangsstadien dieses Institut zu leiten. Die zuständigen Gremien der DFG befürworteten dieses Vorhaben zwar mit Nachdruck, signalisierten aber gleichzeitig die Möglichkeit, daß NRW finanziere.

Bei der Überwindung der Hürde im Kultusministerium half neben dem Rektor Kauffmann und dem Kanzler Schneider der Staatssekretär im Wirtschaftsministerium von NRW, Professor Leo Brand, der unermüdliche und unersetzliche Motor zur Forschungsförderung in NRW. Brand erklärte mir nach einem Besuch, zu dem er Max eingeladen hatte, er fühle sich verpflichtet, sich für die Durchführung des Projektes ganz einzusetzen, da er – übrigens ein Sozialdemokrat, wie man sich ihn wüscht – in der Mitwirkung dieses deutschen Gelehrten die Garantie für das Gelingen einer äußerst wichtigen Forschungseinrichtung sehe. Nach der Zustimmung des Kultusministeriums wurde das geplante Institut ein Universitätsinstitut und dafür war auch das Placet der Fakultät nötig. Hier konnte Max nicht mitwirken. Es entbrannte eine heftige Diskussion. Daß die Fakultät schließlich „ja" sagte, war vor allem der Befürwortung durch den damaligen Chemiker Kurt Alder (Dien-synthese), Nobelpreisträger für Chemie 1952, zu verdanken.

Hier endet meine Erzählung. Wie es in Köln weiterging, wissen Sie. Das Institut wurde gebaut und entfaltete seine Tätigkeit. Zwei der ersten Jahre war es von Max Delbrück geführt. Wir hätten Max und seine Familie gerne für immer hier gehabt. Das war sehr viel, zu viel verlangt. Eine Zeitlang gab es allerdings ein paar Anzeichen, die hoffnungsvoll stimmten. Einmal unterhielt sich Max mit dem Institutsarchitekten Hanns Koerfer einen Abend lang über den Bau eines Privathauses. Vor allem aber hörten wir mal nach 1956, Max rede seine Hilfskräfte in Pasadena nur noch deutsch an! Die Erfüllung unseres Wunsches blieb uns schließlich versagt. Dafür erfreuten uns Max und Manny durch ihren regelmäßigen Besuch bei Deutschlandreisen und das traditionelle Tennisspiel von Max mit unserer Tochter Inge. Im August 1980 erhielt ich von Max einen langen Brief, in dem er von den derzeitigen Reiseplänen seiner Familie und seinen eigenen, eingeschränkten, die nach Cold Spring Harbor gehen, erzählt. Und dann schreibt er: „Also, was gibt es sonst noch Lustiges zu berichten? Jonathan ist in Alaska und verdient ganz schön als Privatfluglehrer. Es gibt dort Scharen von Leuten, die sich ein $ 20.000 — Flugzeug kaufen und dann erst ans Fliegen lernen denken! Ludina übt ihr Cello mit ungeheurem Eifer und Erfolg, von früh bis spät. Tobi ist immer noch mehr an Computern als an Kühen interessiert. Schließlich am 17.August, Sonntag abend, werde ich auf Euren Bildschirmen erscheinen, als „Zeuge des Jahrhunderts" (wer wäre das nicht!) von Peter von Zahn interviewed. Allerhand, was? Gott sei dank kann man's hier nicht sehen". Über diesem Brief, der sein letzter an mich war, lag eine verhalten-heitere Stimmung, so, wie wir Max in unserer Erinnerung bewahren werden.

7

Niels Bohr's Last Lecture — An Introduction[72]

Gunther S. Stent

Thirty years after his seminal "Light and Life" lecture in Copenhagen and a few months before his death in November 1962, Bohr spoke at the dedication ceremony of the newly founded Institute of Genetics at the University of Cologne. Bohr entitled his address, which was to be his last public lecture, "Light and Life Revisited". Aware of the explosive progress in molecular biology that had begun to take place in the meanwhile, he had reconsidered his onetime conjecture about the impossibility of reducing physiology to physics. Bohr now said:

> "It appeared for a long time that the regulatory functions in living organisms, disclosed especially by studies of cell physiology and embryology, exhibited a fineness so unfamiliar to ordinary physical and chemical experience as to point to the existence of fundamental biological laws without counterpart in the properties of inanimate matter studied under simple reproducible experimental conditions. Stressing the difficulties of keeping the organisms alive under conditions which aim at a full atomic account I therefore suggested that the very existence of life might be taken as a basic fact in biology in the same sense as the quantum of

[72]Gunther S. Stent. *Paradoxes of free will.* Transactions of the American Philosophical Society, Vol. 92, Part 6, Philadelphia 2002, pp. 259–260.

> action is to be regarded in atomic physics as a fundamental element irreducible to classical physical concepts."

From the point of view of physics, the mysteries of life were indeed stark. Physiologists had discovered innumerable ways in which cells respond intelligently to changed environmental conditions. Embryologists had demonstrated the possibilities of such amazing feats as growing two whole animals from one embryo split into halves. The transgenerational stability of the gene and the algebra of Mendelian genetics suggested to Bohr that the process underlying the phenomena were akin to quantum mechanics. The resistance of biologists to such ideas did not surprise Bohr. He had met resistance to the complementarity argument before among his fellow physicists.

But this time Bohr was wrong, because James Watson's and Francis Crick's discovery of the DNA double helix in 1953 did for biology what many physicists had hoped, in vain, could be done for atomic physics. It solved all the mysteries in terms of visualizable theories, without having to abandon our intuitive notions about truth and reality embedded in our natural Kantian *a priori* categories of theoretical reason.

8

Light and Life Revisited[73]

Niels Bohr

It is a great pleasure for me to follow the invitation of my old friend Max Delbrück to speak at the inauguration of this new Institute of Genetics at the University of Cologne. Of course, as a physicist I have no first-hand knowledge of the extensive and rapidly developing field of research to which this Institute is devoted, but I welcome Delbrück's suggestion to comment upon some general considerations about the relationship between biology and atomic physics, which I presented in an address entitled "Light and Life", delivered at an International Congress on Radiation Therapy in Copenhagen thirty years ago.

[73]Reprint of Niels Bohr, *Essays 1958–1962 on atomic physics and human knowledge* (New York, 1963), pp. 23–29. Unfinished manuscript. In the preface (pp. vi–vii) Bohr's son, Aage Bohr, wrote: "The two following papers enter on the relationship between physics and biology, which through the years has deeply interested my father and which he first discussed in the address "Light and Life" from 1932, included in the previous volume. He felt that some of his remarks from that time had not always been properly interpreted, and he was anxious to give an account of his views as they had developed since then, in particular under the stimulation of the great discoveries in the field of molecular biology, which he had followed with such enthusiasm.

The subject is touched upon in the third article, written in 1960, but my father was hoping to take up the question in a more detailed paper based on a lecture he delivered in Cologne, in June 1962, with the title "Light and Life Revisited". However, shortly afterwards, he became ill, and although he was well on his way to recovery and had resumed work on the article, it was not completed when he died, suddenly, on November 18th, 1962.

Delbrück, who at that time was working with us in Copenhagen as a physicist, took great interest in such considerations which, as he has been kind enough to say, stimulated his interest in biology and presented him with a challenge in his successful researches in genetics.

The place of living organisms within general physical experience has through the ages attracted the attention of scientists and philosophers. Thus, the integrity of the organisms was felt by Aristotle to present a fundamental difficulty for the assumption of a limited divisibility of matter, in which the school of atomists sought a basis for the understanding of the order reigning in nature in spite of the variety of physical phenomena. Conversely, Lucretius, summing up the arguments for atomic theory, interpreted the growth of a plant from its seed as evidence for the permanence of some elementary structure during the development, a consideration strikingly reminiscent of the approach in modern genetics.

Still, after the development of classical mechanics in the Renaissance and its subsequent fruitful application to the atomistic interpretation of the laws of thermodynamics, the upholding of order in the complicated structure and functions of the organisms was often thought to present unsurmountable difficulties. A new background for the attitude towards such problems was, however, created by the discovery of the quantum of action in the first year of our century, which revealed a feature of individuality in atomic processes going far

---

He left a manuscript which he had prepared as a basis for the lecture in Cologne and in which his views on the relationship between physics and biology are expounded in somewhat greater detail than in his earlier articles. It is, however, only with considerable hesitation that this manuscript is being published. Those familiar with my father's way of working will know what great efforts he devoted to the preparation of all his publications. The text would always be re-written many times while the matter was being gradually elucidated, and until a proper balance was achieved in the presentation of its various aspects. Although a great deal of work had been done in the present manuscript, it was still far from completion. It has nevertheless been included in the present volume, because of the interest which attaches to the views it contains, but the reader must bear in mind the preliminary character of their formulation. With regard to a few passages containing comments on specific biological problems, the author had planned major revisions. These passages have therefore been omitted from the text; their substance is indicated in notes inserted in small print."

beyond the ancient doctrine of the limited divisibility of matter. Indeed, this discovery provided a clue to the remarkable stability of atomic and molecular systems on which the properties of the substances composing our tools as well as our bodies ultimately depend.

The considerations in my address referred to were inspired by the recent completion of a logically consistent formalism of quantum mechanics. This development has essentially clarified the conditions for an objective account in atomic physics, involving the elimination of all subjective judgement. The crucial point is that, even though we have to do with phenomena outside the grasp of a deterministic pictorial description, we must employ common language, suitably refined by the terminology of classical physics, to communicate what we have done and what we have learned by putting questions to nature in the form of experiments. In actual physical experimentation this requirement is fulfilled by using as measuring instruments rigid bodies such as diaphragms, lenses, and photographic plates sufficiently large and heavy to allow an account of their shape and relative positions and displacements without regard to any quantum features inherently involved in their atomic constitution.

In classical physics we assume that phenomena can be subdivided without limit, and that especially the interaction between the measuring instruments and the object under investigation can be disregarded, or at any rate, compensated for. However, the feature of individuality in atomic processes, represented by the universal quantum of action, implies that in quantum physics this interaction is an integral part of the phenomena, for which no separate account can be given if the instruments shall serve their purpose of defining the experimental arrangement and the recording of the observations. The circumstance that such recordings, like the spot produced on a photographic plate by the impact of an electron, involve essentially irreversible processes presents no special difficulty for the interpretation of the experiments, but rather stresses the irreversibility which is implied in principle in the very concept of observation.

The fact that in one and the same well-defined experimental arrangement we generally obtain recordings of different individual processes thus makes indispensable the recourse to a statistical account of quantum phenomena. Moreover, the impossibility of combining phenomena observed under different experimental arrangements into a single classical picture implies that such apparently contradictory phenomena must be regarded as complementary in the sense that, taken together, they exhaust all well-defined knowledge about the atomic objects. Indeed, any logical contradiction in these respects is excluded by the mathematical consistency of the formalism of quantum mechanics, which serves to express the statistical laws holding for observations made under any given set of experimental conditions.

For our theme it is of decisive importance that the fundamental feature of complementarity in quantum physics, adapted as it is to the clarification of the well-known paradoxes concerning the dual character of electromagnetic radiation and material particles, is equally conspicuous in the account of the properties of atomic and molecular systems. Thus, any attempt at space-time location of the electrons in atoms and molecules would demand an experimental arrangement prohibiting the appearance of spectral regularities and chemical bonds. Still, the fact that the atomic nuclei are very much heavier than the electrons allows the fixation of the relative positions of the atoms within molecular structures to an extent sufficient to give concrete significance to the structural formulae which have proved so fruitful in chemical research. Indeed, renouncing pictorial description of the electronic constitution of the atomic systems, and only making use of empirical knowledge of threshold and binding energies in molecular processes, we can within a wide field of experience treat the reactions of such systems by ordinary chemical kinetics, based on the well-established laws of thermodynamics.

These remarks apply not least to biophysics and biochemistry, in which in our century we have witnessed such extraordinary progress. Of course, the practically uniform temperature within the organisms reduces the thermodynamical requirements to constancy or steady decrease of free energy. Thus, the assumption suggests itself that the formation of all permanently or temporarily present macromolecular structures represents essentially irreversible processes which increase the stability of the organism under the prevailing conditions kept up by nutrition and respiration. Also the photo-synthesis in plants is of course, as recently discussed by Britten and Gamow, accompanied by an overall increase in entropy.

Notwithstanding such general considerations, it appeared for a long time that the regulatory functions in living organisms, disclosed especially by studies of cell physiology and embryology, exhibited a fineness so unfamiliar to ordinary physical and chemical experience as to point to the existence of fundamental biological laws without counterpart in the properties of inanimate matter studied under simple reproducible experimental conditions. Stressing the difficulties of keeping the organisms alive under conditions which aim at a full atomic account, I therefore suggested that the very existence of life might be taken as a basic fact in biology, in the same sense as the quantum of action has to be regarded in atomic physics as a fundamental element irreducible to classical physical concepts.

In reconsidering this conjecture from our present standpoint, it must be kept in mind that the task of biology cannot be that of accounting for the fate of each of the innumerable atoms permanently or temporarily included in a living organism. In the study regulatory biological mechanisms the situation is rather that no sharp distinction can be made between the detailed construction of these mechanisms and the functions they fulfil in upholding the life of the whole organism. Indeed, many terms used in practical physiology reflect a procedure of research in which, starting from the recognition of the functional role of the parts of the organism, one aims at a physical and chemical account of their finer structures and of the processes in which they are involved. Surely, as long as for practical or epistemological reasons one speaks of life, such teleological terms will be used in complementing the terminology of molecular biology. This circumstance, however, does not in itself imply any limitation in the application to biology of the well-established principles of atomic physics.[74]

---

[74] In the lecture in Cologne (which was given in German) the author inserted the following phrase: "In the last resort, it is a matter of how one makes headway in biology. I think that the feeling of wonder which the physicists had thirty years ago has taken a new turn. Life will always be a wonder, but what changes is the balance between the feeling of wonder and the courage to try to understand." (Translated from the transcript of the tape recording.)

To approach this fundamental question it is essential to distinguish between separate atomic processes taking place within small spatial extensions and completed within short time intervals, and the constitution and functions of larger structures formed by the agglomeration of molecules keeping together for periods comparable to or exceeding the cycle of cell division. Even such structural elements of the organism often display properties and a behaviour which imply an organization of a more specific kind than that exhibited by the parts of any machine we are able to construct. Indeed, the functions of the building blocks of modern mechanical and electromagnetic calculation devices are determined simply by their shape and by such ordinary material properties as mechanical rigidity, electric conductivity and magnetic susceptibility. As far as the construction of machines is concerned, such materials are formed once and for all by more or less regular crystalline accumulations of atoms, while in the living organisms we have to do with a remarkable rhythm of another kind where molecular polymerisation which, when carried on indefinitely, would make the organism as dead as a crystal, is time and again interrupted.

[A paragraph is here omitted, commenting on the isotopic tracer investigations by Hevesy, which showed that a major part of the calcium atoms incorporated in the skeleton of a mouse at the foetal stage remains there for the whole life of the animal. The author discussed the problem of how the organism is able to economize with its calcium to such a remarkable extent during the growth of the skeleton.]

The application of physical methods and viewpoints has led to great progress in many other fields of biology. Impressive examples are the recent discoveries of the fine structure of muscles and of the transport of the materials used for the activity of the nerves. At the same time as these discoveries add to our knowledge of the complexity of the organisms, they point to possibilities of physical mechanisms which hitherto have escaped notice. In genetics, the early studies by Timofjeev-Ressofskij, Zimmer and Delbrück of the mutations produced by penetrating radiation permitted the first approximate evaluation of the spatial extensions within the chromosomes critical for the stability of the genes. A turning-point in this whole field came, however, about ten years ago with Crick and Watson's ingenious proposal for an interpretation of the structure of the DNA molecules. I vividly remember how Delbrück, in telling me about the discovery, said that it might lead to a revolution in microbiology comparable with the development of atomic physics, initiated by Rutherford's nuclear model of the atom.

In this connection I may also recall how Christian Anfinsen, in his lecture at a symposium in Copenhagen a few years ago, started by saying that he and his colleagues had hitherto considered themselves learned geneticists and biochemists, but that now they felt like amateurs trying to make head and tail of more or less separated biochemical evidence. The situation he pictured was, indeed, strikingly similar to that which confronted physicists by the discovery of the atomic nucleus, which to so unsuspected a degree completed our knowledge about the structure of the atom, challenging us to find out how it could be used for ordering the accumulated information about the physical and chemical properties of matter. As is well known, this goal was largely achieved within a few decades by the cooperation of a whole generation of physicists, which in intensity and scope resembles that taking place these years in genetics and molecular biology.

[A section is here omitted, commenting on the problem of the rhythm in the process of growth of a cell. The author in particular discussed the control of DNA duplication and the role which the structure of the chromosomes may play in this process, as well as in the stability of the genetic material. He further considered the possibility that the duplication process is intimately associated with the transfer of information from DNA.]

Before I conclude, I should like briefly to call attention to the source of biological knowledge which the so-called psychical experience connected with life may offer. I need hardly stress that the word consciousness presents itself in the description of a behaviour so complicated that its communication implies reference to the individual organism's awareness of itself. Moreover, words like "thoughts" and "sentiments" refer to mutually exclusive experiences and have therefore, since the origin of human language, been used in a typically complementary manner. Of course, in objective physical description no reference is made to the observing subject, while in speaking of conscious experience we say "I think" or "I feel". The analogy to the demand of taking all essential features of the experimental arrangement into account in quantum physics is, however, reflected by the different verbs we attach to the pronoun.

The fact that every thing which has come into our consciousness is remembered points to its leaving permanent marks in the organism. Of course we are only here concerned with novel experiences of importance for action or contemplation. Thus, we are normally unconscious of our respiration and the beating of the heart and hardly aware of the working of our muscles and bones during the motion of our limbs. However, by the reception of sense impression on which we act at the moment or later, some irreversible modification occurs in the nervous system, resulting in a new adjustment. Without entering on any more or less naïve picture of the localization and integration of the activity of the brain, it is tempting to compare such adjustment to irreversible processes by which stability in the novel situation is restored. Of course, only the possibility of such processes, but not their actual traces, are hereditary, leaving coming generations unencumbered by the history of thinking, however valuable it may be for their education.

In expressing the warmest wishes for the success of the investigations of the distinguished group of scientists working in this new and magnificently equipped Institute, I cannot think of a better prospect than that it will contribute to increase our insight into that order of nature which it was the original aim of the atomic conception to account for.

III. The Beginnings

9

Working with Max Delbrück

Charles N. David

It's nice to be back at the Institute of Genetics in Cologne. After the first four talks this morning, I think we are now in for a little lighter fare. At least that's what I have in mind and I suspect some of the other speakers this afternoon are going to give you a bit less history and few more entertaining details.

The organisers invited me to talk about working with Max Delbrück and I want to divide my comments into two parts: "Not for the faint hearted" and "A family affair". You will realise what these two parts mean in a few minutes. I was a graduate student with Max Delbrück in the 1960s. I spent one year in Cologne in the new Institute of Genetics from the fall of 1962 to the summer of 1963 and then four years with Delbrück in Pasadena. Delbrück had a very profound influence on my life, both scientifically and, as you will see, personally.

Not for the faint hearted

Max Delbrück worked with a great many people in his long and very diverse scientific career: Timofeeff-Ressovsky was mentioned earlier; Salvador Luria got the Nobel Prize with Max in 1969; Jean Weigle was his long time colleague at Caltech; Werner Reichardt was his *Seelenkollege* in Tübingen; Gunther Stent was a post-doc and then long-time friend of Delbrück in Berkeley; Jim Watson you all know;

and Peter Starlinger you just met. All these people are mentioned in the very nice biography of Max Delbrück by Ernst Peter Fischer. But I am not mentioned, so you could begin to wonder if I ever worked with Max Delbrück. Yes, I did and the picture in Fig. 7 will hopefully prove it. It was taken on one of the regular Max Delbrück *Spaziergänge* in Königsforst in 1962, shortly after I arrived in Cologne. It shows the two of us and it was the beginning of our five years together: Max as professor and I as a graduate student. The names on the picture are Max and Carlos. That is because I came to Cologne by the long route. I drove a Landrover from Boston through Central and South America to Bolivia, hitchhiked to Brazil, took a boat to Europe and finally got to Cologne several months later than Max had expected me. He never commented on this, but proceeded to call me Carlos for the rest of his life. The name has stuck. My wife also calls me Carlos.

Fig. 7. Max and Carlos, Königsforst, Köln, 1962.

When I came to Cologne, Max was working on UV-damage in DNA and he convinced me to take up this topic. It had just been shown by the Setlows that UV irradiation of DNA induced thymine dimerisation and most people were pretty convinced thymine dimers constituted lethal hits in DNA. Work in several labs had shown that the single stranded DNA virus ΦX174 was dramatically more sensitive to UV than the double stranded virus T4. Max was quite concerned about this, since it looked as if the quantum yield for thymine dimerisation might be different in different forms of DNA. Hence, one of the projects in the lab was to find out whether the quantum yield was in fact the same or different.

A post-doc, Danny Wulff, developed very sensitive methods for measuring thymine dimers in DNA and proceeded to determine the quantum yield for dimer formation in double stranded DNA. Max got me to do the same experiment with single stranded DNA. It turned out that the answers were the same, which was very encouraging for Max as a physicist. But these experiments also revealed that there were only 0.34 dimers per lethal hit in ΦX174. This indicated that there must be other UV lesions in DNA, which at the time wasn't what most people, including Max, expected. This was a pretty nice piece of work for a beginning graduate student and, as you can see in Fig. 8, it was nicely published (in *Zeitschrift für Vererbungslehre*). You can also see that Max Delbrück is not an author. The reason is simple: Max was not interested any more! This outcome was a bit disturbing for me and is the reason for my comment at the beginning "not for the faint hearted".

After a year in Cologne working on thymine dimerisation and UV lesions Max went back to Pasadena and I went with him. When we got to Pasadena, Max more or less immediately forgot thymine dimers and went back to work on the project which he had started in the 1950s and which would occupy him for the rest of his scientific career, namely how sensory transduction takes place in biological systems. It has become a very common topic these days but in the 1950s and the 1960s almost nobody had heard of the problem of sensory transduction. Max was once again pioneering a new field: in this case, how light is converted into a biological response. The

Division of Biology, California Institute of Technology Pasadena

UV INACTIVATION AND THYMINE DIMERIZATION IN BACTERIOPHAGE ΦX

By

CHARLES N. DAVID

With 3 Figures in the Text

(Received October 15, 1964)

Summary

Thymine dimers are found in DNA following irradiation at 260 mμ. The quantum yield for dimer formation in bacteriophage ΦX (single-stranded DNA) is 0.013 dimers per quantum absorbed by a nucleotide. This is comparable to the quantum yield for bacteriophage T4v_1 (double-stranded DNA) indicating there is no dependence of thymine dimerization on the nature of the irradiated DNA.

In ΦX the number of thymine dimers per lethal hit is 0.34. This demonstrates the existence of other as yet unidentified lethal photoproducts in irradiated ΦX DNA.

Acknowledgements: The author wishes to thank M. DELBRÜCK, D. WULFF, and W. SAUERBIER for advice and help in this work.

Fig. 8. Section of a paper by Charles N. David, in *Zeitschrift für Vererbungslehre* in 1964.

organism he picked, *Phycomyces*, is a fungus which grows as a mycelial mat but puts up aerial sporangiophores that elongate extremely rapidly. After an initial elongation phase the sporangiophore forms a ball of spores and then continues elongating. What interested Max was that sporangiophores are exquisitely sensitive to light. If they are illuminated unilaterally, they grow towards the light, i.e. they are phototropic. What happens inside the sporangiophore, when it senses light and grows toward the light source? In typical Max fashion, he got a post-doc, Hyde, in the EM [electron microscope] laboratory at Caltech to make EM sections of sporangiophores, in particular of the grow zone. The result is in Fig. 9.

In thin sections of the grow zone Max and Hyde discovered subcellular particles which Max had never seen before and which actually nobody else had ever seen before. Max immediately dubbed them "Hyde particles" after the man who helped him with the experiments. Hyde particles were small and round with a two dimensional

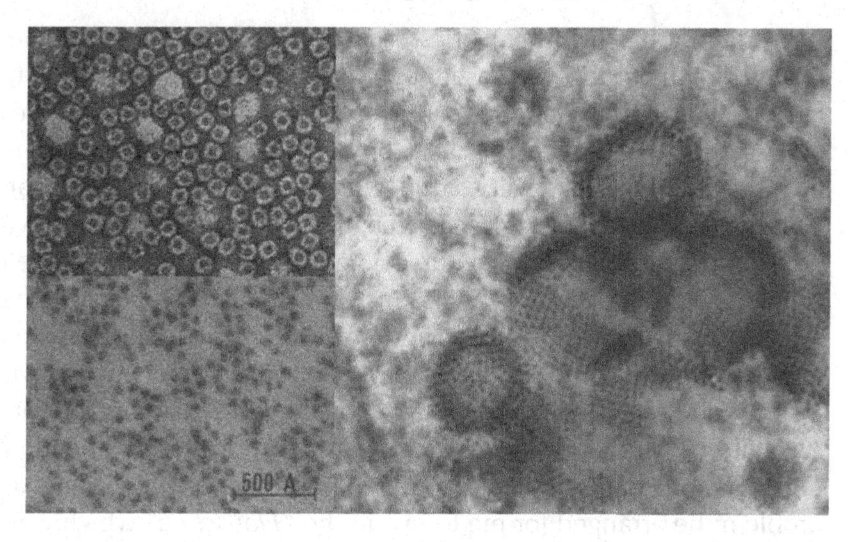

Fig. 9. Phycomyces ferritin or the "Hyde particles" in the growth zone.

grid of dark spots on them and they were uniquely present in the grow zone, which is light sensitive. So Max was convinced that Hyde particles were the eyes of sporangiophores!

I was still an impressionable graduate student in those days, and so I took Max's advice and went to work on Hyde particles. The answer is shown in Fig. 9: Hyde particles are lipid droplets covered with a two dimensional crystalline array of ferritin molecules. This was a pretty uninteresting result, if you are trying to find eyes. That is also what Max thought! As I mentioned previously: being a graduate student with Max was "not for the faint hearted".

We now know that ferritin is one of the most conserved molecules in biology. It is present in plants, animals and fungi. *Phycomyces* ferritin, however, was the first non-mammalian ferritin to be discovered, and I was pretty proud of that discovery. However, since Max was no longer interested, I wrote up my thesis and graduated. The publication in the *Journal of Cell Biology* describing Hyde particles and *Phycomyces* ferritin did not list Max Delbrück as an author.

Everybody has his Max stories. I have lots of them, one of which relates directly to the new Institute in Cologne. When I arrived in the fall of 1962, I spoke no German. Max, however, had established a policy that only German could be spoken in the Institute. He banned English, even though every young German graduate student and post-doc was desperate to speak English with the Americans who were floating through the Institute. When we met for the first time in Cologne, Max spoke English with me for an hour. Then he switched into German. I did not understand a word and it took me several months to get over the language barrier. To make matters worse, Max insisted on correcting every mistake I made while learning German. As I said before, working with Max was "not for the faint hearted". On the other hand Max also provided a solution to the language problem: he arranged for me to live in the "*Häuschen*" which was a small house next to the Institute where three graduate students already lived. They were all Germans and we spoke only German. As a result I learned German very rapidly, mostly evenings in the local bar with these students.

A family affair

I want to turn now to the second half of this brief summary about working with Max Delbrück: a family affair. Max was a family man. Joining Max's lab was the same as joining the family. And the best part about that was Max's wife Manny. She was the one who picked up the graduate students who were periodically faint hearted.

Joining the Delbrück family meant going on camping trips to the desert in southern California. That was part of the tradition. The basic instructions were: drive to Mecca (a town near the Salton Sea in southern California) and turn left (see Fig. 10). The picture shows a typical hike on one of these trips with a horde of Delbrück friends and graduate students. In the foreground is Carlos with Toby Delbrück [Delbrück's youngest son] and what Toby thought was a sleeping flower.

Fig. 10. Camping in the desert; "drive to Mecca and turn left". Carlos and Toby Delbrück and the sleeping flower.

Finally, I want to mention that working with Max had an important non-scientific consequence for me: I met my wife Almuth Kähler. The Kählers lived right next to the Delbrücks in the Bachemerstrasse just around the corner from the Institute. The Kählers had three daughters of marriageable age and I was a young man of marriageable age. So the Delbrücks, primarily Manny, got me together with them relatively quickly. I promptly fell in love with Almuth, but after that year in Cologne, she went to Berlin to study architecture and I went to California with Max. In those days intercontinental travel was expensive and we had a rather complicated relationship for almost eight years. But we did finally get it together and we got married in Cologne in the Bachemerstrasse in 1970. Max and Manny were there.

10

Recollections

Hans G. Zachau

Professor Zachau kindly sent us extracts from two essays that he wrote shortly after his retirement. We first cite from "Life with tRNA, chromatin, immunoglobulin genes: recollections of a German molecular biologist", *Comprehensive Biochemistry* **41** (2000), pp. 635–666.

Becoming a biochemist

"I am one of those lucky individuals who have witnessed the emergence of molecular biology from early on and have participated in its endeavours. Why do I want to write about that and for whom? The 'why question' is easily answered: autobiographical writing is fun for the writer and it is an appropriate pastime for a retired professor. But who may want to read the article? Perhaps some other scientists; younger or older ones. When I was a student or young scientist I probably would not have looked at what an old professor had written about his life in science. However, I became interested in biographies and books on history after I had reached middle age. Now a good part of my spare time reading is spent with biographies, not only of geniuses, but also of normal people, scientists and others. I like to read how people, ideas and situations develop. Maybe some of my colleagues are similarly inclined and take an interest in my recollections."

Hans Zachau who as a pupil attended a lecture by Max Delbrück at Berlin-Dahlem in 1948, studied medicine and chemistry and decided to become a biochemist — that is, at the time, a physiological chemist. Most influential in his decision was his reading of many popular or semi-popular books on medicine and various fields of science. In reading these, he began to feel that the most interesting things in the coming next decades would lie somewhere between medicine and chemistry. He conducted biochemical research for his diploma and Ph.D. theses at the MPI für Biochemie in Tübingen under Adolf Butenandt, and received his Ph.D. in chemistry in 1955.

"I still had not touched real biochemistry or molecular biology and was not even sure what it would be like. I wanted to learn this in the US, and a visiting lecturer in Tübingen, Professor J.C. Sheehan of MIT, convinced me that his kind of peptide chemistry was the right thing for me to do. In the one year in Tübingen before I went to the US I was given two assignments: to study the pheromones of the honey bee and to isolate DNA from pine pollen. The first topic was related to my thesis work and the second one came about because Butenandt knew of my interest in nucleic acids. I did not find anything startling and was eventually happy to leave. Butenandt asked me to promise to him by handshake that I would come back to Germany, and offered that I could return to his institute. I received a Fulbright travel grant, which made me a 'half bright', since unlike a true Fulbright fellow I got only travel money. My salary was to come from MIT. The exam of the Fulbright Commission was for me the first occasion to memorize the American presidents in sequence and all states of the Union."

[...]

"I still had in mind to work on nucleic acids some day and when, during my time at MIT, I heard a lecture by Fritz Lipmann I knew that after finishing the etamycin structure I wanted to work with him. Lipmann was hesitant at first, which in turn offended Butenandt, who wrote me that there were several places in the US where people from his institute were welcome. But I knew about these places from colleagues at Butenandt's institute who had been there for their post-doctoral, and I was not attracted by the work done there. Clearly many scientists in the US, probably also Lipmann, resented the fact that Butenandt had taken the directorship of the Kaiser-Wilhelm-Institut für Biochemie in Berlin from Carl Neuberg, who was forced to emigrate as a Jew. My problem with Lipmann was solved, also to the eventual satisfaction of Butenandt, with the help of Gerhard Schmidt, known for the Schmidt-Thannhauser reaction. He was another emigrant from Germany in Boston who knew Lipmann well and, indirectly, also my parents. So I was accepted by Lipmann for 1957/58, the first year after his moving from Boston to the Rockefeller University in New York."

[...]

"The work in Lipmann's lab was what I had hoped for: mechanistic studies on an enzyme which later was called tryptophenyl tRNA synthetase, and a search for the site of amino acid attachment to the soluble RNA fraction, later named tRNA. The outcome of the second project, conducted together with G. Acs, a Hungarian post-doc in the lab, was that the amino acids are esterified to the 2′ or 3′ hydroxyl group of the 5′ terminal adenosine of tRNA, my first finding which entered the textbooks.[75] It also led to a short talk, which I was allowed to give at a Gordon Conference. This series of meetings was and is devoted to quite specialised topics, but it was characteristic for the state of molecular biology in 1958 that this meeting still was a 'Gordon Conference on Proteins and Nucleic Acids'."

After the two post-doctoral years in the United States, Hans Zachau returned to Germany in 1958, where Butenandt had reserved him a place in his MPI.

[75]H.G. Zachau, G. Acs, F. Lipmann, "Isolation of adenosine amino acids esters from a ribo-nuclease digest of soluble, liver ribonucleic acid", *Proc. Natl. Acad. Sci. USA* **44** (1958), pp. 885–889.

tRNA research in Munich and Cologne, 1958–66

"I was lucky to get a nice lab in Butenandt's institute, which had moved from Tübingen to Munich in 1956/57. During the first two years I assembled a small group: a technician, a Ph.D. and a diploma student, two Japanese post-docs, and Bill Lawson who came for a year as a visiting scientist. Being embedded in the Max Planck Institute, I did not have to worry about salaries or fellowships for the co-workers or money for chemicals. In Lipmann's lab I had fallen in love with tRNA and Lipmann had agreed to my plans of continuing work with tRNA on my own. We developed a cheap, large scale procedure to isolate tRNA from brewer's yeast with technical grade phenol. Much time was spent on more or less fancy methods of fractionation aimed at the isolation of a pure tRNA species suitable for structural work. Eventually counter-current distribution of amine salts of the RNA proved to be the most reliable method. It yielded pure serine specific tRNA. Nuclease digestion permitted a first glimpse at its structure. Besides the 'risky' tRNA work, 'safe' projects were carried out with the students. They dealt with amino acid esters of nucleosides, sugars and model compounds aiming at a stereochemical explanation for the high reactivity of the tRNA-bound amino acid. Fortunately we did not have to follow these projects for long, since the tRNA work came along well."

"In 1960 Max Delbrück looked for a biochemist to establish a group in his future genetics institute in Cologne. He also asked Butenandt who referred him to several co-workers, including me. We had a long talk, but, before making an offer, Delbrück wanted to meet my (newly wed) wife. They got along fine and a few weeks later it was decided that we would move to Cologne in the summer of 1961. My first step on the academic ladder, the *Habilitation*, was arranged between Professors Butenandt, Delbrück and Klenk. It was a *Habilitation* in physiological chemistry, a term used in medical faculties (see above) for a person working in an institute of the science faculty; it took place in Cologne with work mostly done in Munich. Reasonable people already considered the *Habilitation* at that time to be an outdated but (in Germany) unavoidable ritual. I did not have to teach physiological chemistry to medical students but participated in the teaching of the genetics institute. I only vaguely remember the difficulties of getting work started with a new group of people in a not-quite-finished building. What I do remember, however, is the contrast between Delbrück's non-hierarchical enclave in a German-style faculty and university. The genetics institute attracted German sympathisers and, of course, many foreign visitors. It became known for its lively scientific atmosphere, for the phage courses, which Delbrück organised, and last but not least, for its parties, not only at Karneval."

"The work on tRNA isolation was time consuming and the subsequent structural work presented tough problems. Much effort had to be invested into developing methods of what today is called RNA sequencing. One of the methods, although unsuccessful, became important for us, since it provided, for a while, most of our funds for research. Different from the situation in Munich, I now had to write grant applications, although Max Delbrück divided the resources of the Institute equally between himself and the four groups of the Institute. The method I am referring to made use of the powerful UV monochromator which Delbrück built in Cologne. The idea was to create UU dimers in tRNA by irradiation at 280 nm; digest the product with nucleases; isolate the UU-containing fragments; cleave them at the UU sites by irradiation with 240 nm light; and identify the oligonucleotide products which must have been adjacent in the original sequence. The method worked only 'in principle', but irradiation of nucleic acids had a high priority in the funding policy of the German research ministry. Several methods had to be combined in the final attack on the tRNA structure. One of the more important ones involved sequential cycles of partial endonuclease digestion, isolation of fragments, and recleavage with exo- and endonucleases. In 1965 R.W. Holley elucidated the structure of alanine tRNA, and we were second, completing the work on serine tRNA in 1966. D. Dütting, H. Feldmann and, at an earlier stage, F. Melchers were the colleagues who contributed most to the structural work.[76] We had, in

[76]H.G. Zachau, D. Dütting, H. Feldmann, "The structures of two serine transfer ribonucleic acids", *Hoppe-Seyler's Z. Physiol. Chem. 347* (1966), pp. 212–235.

fact, analyzed two serine tRNAs, termed 1 and 2, which differed in three of the 85 nucleotide positions. The gene for the serine tRNA 1 was not found in the yeast strain used later for genome sequencing; therefore, this tRNA may be particular to the brewer's yeast we had used. The gene for the serine tRNA 2, however, was found in the yeast genome, and our sequence was confirmed at the DNA level. Since the structure of the alanine tRNA of Holley's group had to be corrected in 1975, we can have the little piece of satisfaction that our serine tRNA 2 was the first fully correct sequence of a naturally occurring nucleic acid established anywhere."[77]

"After the *Habilitation* I was now a *Privatdozent.* In this position one has to make oneself known in places where vacancies of professorships may arise; one therefore has to go to local meetings and accept all invitations to lectures. Seeing different universities was not uninteresting and industrial companies were financially generous, but fortunately my time as a '*Privatdozent*' did not last very long. In 1965/66 I received and negotiated offers of chairs of physiological chemistry in Berlin and Munich, and of genetics in Cologne. The decision was in favour of Munich and in early 1967 my family and I, and a small group of co-workers moved there."

From the recollections that Hans Zachau wrote for his family ("Drei Generationen der Familie Zachau, beschrieben von Hans Georg Zachau", Fassung vom 20. November 2001), we add here some personal comments about his work in Cologne, the special atmosphere at the Institute of Genetics, and the many parties.

„Ein besonderes Fest gab es zur Institutseinweihung im Juni 1962. Der greise Physiker Niels Bohr, ein Lehrer von Max Delbrück, hielt einen praktisch unverständlichen Vortrag über erkenntnistheoretische Probleme der Physik und Biologie. Max hat uns später, als der Festestrubel verrauscht war, den Inhalt mit Erklärungen wiedergegeben.

[77]For a survey of yeast tRNA genes see the homepage of H. Feldmann: *http://www.med.uni-muenchen.de/biochemie/feldmann/*; see pages Research.htm or trna_ty.htm and SPETSAI.html.

Ein weiteres Fest gab es am Ende des gleichen Jahres, als Jim Watson von der Verleihung des Nobelpreises an ihn in Stockholm direkt nach Köln kam, um seinen Lehrer Delbrück zu besuchen.

Damit Delbrück nicht nur im Zusammenhang mit Festen und Geselligkeit erwähnt wird, will ich den UV-Monochromator beschreiben, den er mit Hilfe der Werkstatt im Kölner Institut baute. Das Gerät war für die Strahlenbiologen des Instituts bestimmt und vor allem für seine eigenen Arbeiten mit dem Licht-sensitiven Einzeller *Phycomyces*. Delbrück wollte *Phycomyces* zum 'Bakteriophagen der Sinnesphysiologie' entwickeln, eine Hoffnung, die sich meines Wissens bisher nicht erfüllt hat. Der Monochromator war größer als ein Konzertflügel und hatte als Kernstück ein großes mit Wasser gefülltes Quarzprisma. Als ich mich aus Gründen, die in den 'recollections' beschrieben sind, dafür interessierte, tRNA konsekutiv mit UV-Licht zweier verschiedener Wellenlängen zu bestrahlen, gab mir Delbrück Nachhilfeunterricht in Optik und Messtechnik und führte auch die ersten Versuche gemeinsam mit mir durch."

[...]

„Im Sommer 1963 kehrte Delbrück in die USA zurück. Er wurde natürlich mit einem großen Fest verabschiedet. Er kam häufiger zurück nach Deutschland, nach Köln, zu uns nach München und auch nach Konstanz. 1969 erhielt er den Nobelpreis für Medizin.

Der aufgestauten Animosität etlicher Fakultätskollegen und mancher festgefügter Instituts-Belegschaften in der Nachbarschaft gegen Delbrück und sein Institut waren wir nun ungeschützt ausgesetzt. Vor allem in der näheren Umgebung gab es manchen lästigen Kleinkrieg. Aber sowohl in Deutschland als auch im Ausland hatte das Institut auch viele Sympathisanten, so dass nichts Schlimmes passierte. Wir konnten für Delbrück und zwei der anderen ursprünglichen Abteilungsleiter, die kurz nach ihm das Institut verließen, Nachfolger berufen. Als ich 1967 nach München ging, war Benno Müller-Hill mein Nachfolger im Kölner Institut. Peter Starlinger war der einzige, der die Kontinuität von Anfang an bis zur Mitter der 90iger Jahre aufrechthielt. Es war ein gutes Institut mit vier bis fünf Lehrstühlen entstanden, die sich verschiedenen Zweigen der modernen Biologie widmeten. Dabei kam dem Institut zustatten, dass die moderne Biologie weltweit und schließlich auch in Deutschland boomte, und dass eine Departmentstruktur wie im Institut für Genetik auch andernorts in Deutschland eingeführt wurde. Delbrücks Vision war Realität geworden."

11

How Chemistry Met Genetics

Horst Feldmann

The most remarkable switch in my career occurred when in September 1962 when I was engaged by Hans Zachau (see Fig. 11) to work in his group as a post-doctoral fellow at the Institute of Genetics. This relates to scientific as well as social aspects. What I had experienced thus far was strictly regulated training as an organic chemist. At that time, the Chemical Institute in Cologne was housed in the buildings of the Augusta Hospital in Zülpicher Strasse. Since then, the area has been used to build new facilities, the new Institute of Genetics being one of the latest. During their diploma or even doctoral thesis work in chemistry, the trainees had to fully pay for any equipment and material. Bench space was limited to about 1.5m in width, and normally 40 to 50 people had to share a huge laboratory. Despite this high density, there was little contact or communication among the fellows. Actually, everyone buried himself in his own subject, mostly not realising what his neighbours were engaged in. The only common activity consisted in (rather boring) monthly seminars, where every fellow had to report news he found in periodicals he had been assigned to.

How different life became in the new Institute of Genetics! While in organic chemistry there were only two departments, each employing some three assistants and the same number of technicians, the Institute of Genetics accommodated five departments, each, on average, endowed with six scientific co-workers and roughly the same

Fig. 11. Painting by Rainer Thiebe for a party at the Institute; angel Zachau.

number of technicians. Conceived as an inter-disciplinary research institute, genetics was open to collaborators from different fields, such as physics, chemistry, biology or medicine, and most remarkably, to foreign co-workers. A completely new experience for me was to find a perfect "infrastructure": secretary, workshop, library, clean kitchen, etc.

A drastically new experience for me was team-work; the superior maxim in every department. People talked to each other and collaborated. One of the outstanding topics in discussions at that time (1962/63) was the Genetic Code: triple, quadruple, comma-free or not? Sometimes it took some effort for a beginner in genetics and molecular biology to adapt to the new field. Contact between the people from the single departments was guaranteed by the weekly seminars or colloquia. In the seminars, all researchers and doctoral fellows had to elaborate on a given topic; the colloquia were covered by invited speakers. For 1963 only, I have counted 68 renowned scientists from all over the world visiting the Genetics Institute. (When Max was around, he used to sit in the first row in order to "control" the speaker. If he felt something mysterious in this presentation, he turned to the audience: "Everybody got it?" If there was no answer, the baffled speaker was urged: "Better say it again!")

The relaxed atmosphere became manifest in the many parties (see Fig. 12), for which Max had a foible, and which largely his wife Manny was organising. A most spectacular one was the Farewell Party for Max on July 19, 1963, when he had decided to return to Pasadena. In our sketch, Max was damned to be tied and boiled by the "wild Zachaus" in a huge container, which we used for our large scale tRNA preparations. To mimic the boiling water, we had put some dry ice together with a little water in the bottom. After a while Max sighed in great pain: "Can't you at least remove the dry ice; my back is already burning!" It would, of course, be really tempting to tell more anecdotes about people and life in the Institute during the first years.

Undoubtedly, and most remarkably, the unconventional setup of the Institute of Genetics made it the birthplace for molecular biology in Germany. Remember that around that time — and even for so many years to follow — no training, even in biochemistry, was offered at German Universities. In organic chemistry, protein chemistry was touched only peripherally, and nucleic acid chemistry simply did not exist. So my first notion of this new field stems from a lecture by the late Fritz Cramer. In 1962, as a member of Hans Zachau's team, I could engage myself in this field, experiencing and applying the new techniques in helping decipher the primary structure of the major serine specific tRNAs from yeast (see Fig. 13). This subject challenged the analytical skills of a chemist, but at the same time met my interest in molecular architecture. In fact, it was pure and hard chemistry at the beginning. We had to isolate the starting material from large quantities of brewer's yeast (I guess, during this period I myself worked up about a ton of yeast slurry, which we had to source from

Fig. 12. Paintings of the Institute's members for a Carnival party at the Institute (1963). *Above row*: Lieutenant Harm, Able-bodied seaman Zachau, Vice commander Starlinger. *Below*: Commander Delbrück.

SESTER Kölsch brewery in Cologne), by stirring each batch with 200 l of a phenol/buffer mixture, decanting the aqueous solution, and precipitating the soluble RNA by ca. 250 l ethanol. You will understand that — also with the subsequently developing preparative and

Fig. 13. "3rd floor". The Zachau lab. First (*top*) row: Susanne Notz, Hans Georg Zachau, Fritz Melchers, Dieter Dütting. 2nd row: secretary, Horst Feldmann, Hubert Gottschling. 3rd row: Gisela Schultz, Paula Prüfert, Rainer Thiebe, Wolfgang Karau. Bottom row: Gudrun Patzelt, Heidi Heusinger.

analytical methods — we were far away from the present elegant nano-techniques. But still, there is much of chemistry behind today's molecular biology, so that training in, or knowledge of chemistry remains advantageous if not indispensable.

My first contact with yeast extended to a general topic of my research during my time at the Institute of Physiological Chemistry in Munich. What I had learned in molecular biology from the "early days" on, I tried to pass on to our students, and I am happy that many of the medical students realised the importance of molecular biology for modern medicine.

12

Eindrücke eines Doktoranden

Fritz Melchers

In meinen Erinnerungen an die Zeit als Doktorand im Kölner Institut für Genetik benutze ich die einmalige Gelegenheit, mich aus der Frosch-Perspektive zu erinnern. Ich bin also ganz unschuldig und ohne retrospektive Absicht: ein wirklicher Graduate Student, ein Doktorand des Institutes und nicht einer, der schon weiß, daß er später einmal an präBZell-Rezeptoren im Immunsystem und an der Plastizität von Stammzellen arbeiten wird, und der ein wunderbares wissenschaftlich-menschliches Schicksal hinter sich hat. Ich betrachte mich also einmal für die nächsten zehn Minuten oder Viertelstunde als einen jungen Chemiestudenten, ursprünglich aus Berlin kommend, der hier in Köln Chemie studierte und dann eine Diplomarbeit machte.

Als ich mein Diplom in physiologischer Chemie hier an der Universität erhalten hatte und mir Gedanken gemacht habe, was ich machen sollte, bin ich auf die Idee gekommen, ich sollte vielleicht nach München gehen, wo der Herr Professor Lynen ein aufregendes Labor hatte. Mein Vater [Georg Melchers] hat gesagt: „Du kannst in Gottes Namen nach München gehen, wenn Du unbedingt mußt, aber wenn Du zu Butenandt gehst, dann streiche ich Dir die Bezüge".

Also bin ich im Sommer 1961 nach München gekommen und habe bei Guido Hartmann und der technischen Assistentin Fräulein Coy, die spätere Frau Henning, ein Praktikum gemacht, sozusagen ein Individualpraktikum in Biochemie. Hartmann hat dann zu mir gesagt: „Sagen Sie mal, Herr Melchers, warum kommen Sie denn ausgerechnet nach München? Sind Sie verrückt geworden? Wo doch in Köln so etwas Phantastisches passiert?" Und das wußte ich ja auch schon. Ich hatte ja schon als Student, als Hilfsstudent samstags bei Wolfgang Abel in der näheren Umgebung von Joseph Straub Pflanzenkulturen umgeimpft, und eigentlich wußte ich ja, daß ich wieder hierher kommen sollte.

Also bin ich wieder nach Köln zurückgegangen und habe mich bei Max Delbrück vorgestellt. Ich werde es nie vergessen: Er hatte ein Kalottenmodell vor sich auf dem Tisch liegen und fragte: „Was ist das?" Ich habe geantwortet: „Woher soll ich denn das wissen". Worauf er sagte: „Du bist doch Chemiker, du müßtest das doch wissen!" „Nein", habe ich gesagt, „weiß ich nicht". Das Modell war Thymidin. Daraufhin hat er gesagt: „Na ja, wenn Du Chemiker bist, dann bist Du wohl am besten bei den Biochemikern hier aufgehoben, dann geh' mal zu Zachau und frag' ob er was für Dich zu tun hat". So bin ich dann als Doktorand bei Zachau akzeptiert worden. Der Auszug aus meinem Studienbuch zeigt, daß ich auch unterrichtet worden bin (Fig. 14).

Das Wichtigste an diesem Dokument ist eigentlich, daß von Anfang an die „Anleitung zu wissenschaftlichem Arbeiten" von Max Delbrück unterschrieben wurde. Das hat mir enorm geholfen. Es was also nicht nur die Biochemie von der transfer-RNA und die wirklich exakte und ordentliche Biochemie von Hans-Georg Zachau, sondern auch der zusätzliche Schutz vor dieser Biochemie und die ausweitenden Blicke in den Rest der Molekulargenetik von Max, die mir eine unglaubliche Chance gegeben haben. Ich wurde mit Zachau einig, daß ich transfer-RNA sequenzieren sollte. Er selber arbeitete an dem Problem, ob die Aminosäuren nun an der 2′ oder an der 3′ Position der endständigen Ribose hängen. So begann ich erst einmal alleine, und später zusammen mit Dieter Dütting, an diesem Problem der Sequenzierung zu arbeiten. Um es einfach zusammenzufassen, oder damit es die Leuten wissen, die heutzutage sequenzieren, wir haben für 80 Nukleotide damals drei Jahre gebraucht. Diese Balance während meiner Doktorarbeit zwischen Zachau und Delbrück hatte viele, viele Vorteile. Es war auf der einen Seite die wirklich existierende,

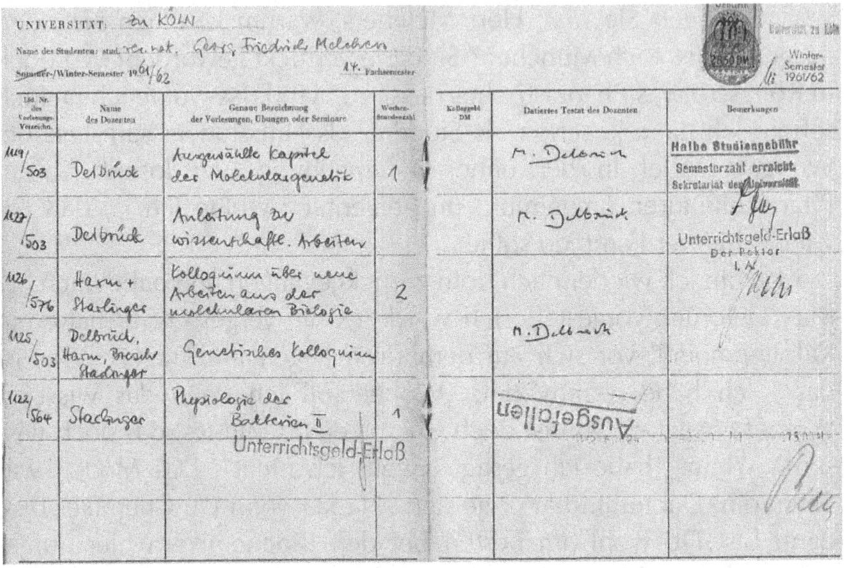

UNIVERSITÄT: zu Köln
Name des Studenten: stud. rer. nat. Georg Friedrich Melchers
Sommer-/Winter-Semester 1961/62 14. Fachsemester
Winter-Semester 1961/62

Lfd. Nr. des Vorlesungs-Verzeichn.	Name des Dozenten	Genaue Bezeichnung der Vorlesungen, Übungen oder Seminare	Wochen-Stundenzahl	Kollegegeld DM	Datiertes Testat des Dozenten	Bemerkungen
1119/503	Delbrück	Ausgewählte Kapitel der Molekulargenetik	1		M. Delbrück	Halbe Studiengebühr Semesterzahl erreicht. Sekretariat der Universität
1127/503	Delbrück	Anleitung zu wissenschaftl. Arbeiten			M. Delbrück	Unterrichtsgeld-Erlaß Der Rektor I. A.
1126/576	Harm + Starlinger	Kolloquium über neue Arbeiten aus der molekularen Biologie	2			
1125/503	Delbrück, Harm, Bresch, Starlinger	Genetisches Kolloquium			M. Delbrück	
1122/564	Starlinger	Physiologie der Bakterien II	1		Ausgefallen	

Unterrichtsgeld-Erlaß

Fig. 14. Record of study of Fritz Melchers.

wie soll ich sagen, die Prägnanz eines Zachaus, der auf die kleinen Details Wert legte, die ein Biochemiker zu beachten hatte und andererseits die oft zitierte fröhliche Respektlosigkeit von Max Delbrück, der Autoritäten immer anzweifelte und der sagte, so richtig kann das auch nicht sein, die richtige Balance, die mir die weitere Sicht in die moderne Genetik gegeben hat.

Im Beitrag von Carsten Bresch sind zwei Beispiele für Max' Auftreten hier in Köln zitiert worden. Ich will dazu noch etwas ergänzen. Das erste betrifft den letzten Tag des Phagen-Kurses. Nicht etwa lagen die Zeitungen auf jedem Sitz, sondern sie lagen auf einem Stuhl am Eingang des Hörsaals und es war jedem freigestellt, sich eine Zeitung zu nehmen, trotzdem hatte am Schluß jeder eine Zeitung genommen und saß lesend da. Es stimmt nicht, daß Max angefangen hat zu reden, sondern er war ausgesprochen beleidigt und sagte: „Ich rede nicht, wenn Ihr nicht die Zeitung wieder weglegt". Das ist ein Beispiel dafür, daß Max' Humor Grenzen hatte.

Das zweite Beispiel ist nun wieder eher schon der Humor, in Ergänzung dessen was Carsten Bresch von Jim Watsons berühmtem

Besuch nach der Erlangung des Nobelpreises hier in Köln erzählt hat. Der 6. Stock war dekoriert mit Leuten, die normalerweise nie dorthin kommen, von Damen mit dicken Hüten und Herren, die sonst normalerweise beim Lehrstuhl für Genetik nie auftauchten. Natürlich hat Jim auf dem Tisch vor der Tafel gelegen und Peter Starlinger hat alles erst auf Englisch herausgeschrieen und dann alles noch mal in deutscher Übersetzung wiederholt. Nach fünf Minuten sagte Max: "Jim, is that all you want to tell me, is that the stuff you told me three months ago?" Da flüsterte Jim noch: "Yes", und daraufhin sagte Max: "Well, if you don't talk about anything else, I'll leave!" Woraufhin Jim Watson seine Sprache wieder fand, aufstand und einen ganz normalen Vortrag hielt.

Ich werde auch nie vergessen, wie Max sich nach zehn Minuten einer Vorlesung oder eines Seminars zur Zuhörerschaft drehte und fragte: "You got this?" Oder er zum Vortragenden sagte: "I haven't understood a word so far. Could you start all over again? And please make three sentences out of one". So war er, er war schrecklich, er war ganz fürchterlich gefürchtet von den Leuten, die hier einen Vortrag halten mußten.

Ein paar Worte zu unserer Arbeit im Labor: Einige Dinge wurden schon erwähnt. Wir haben in diesen riesigen Bottichen transfer RNA gemacht, wir haben Ratten getötet und ihnen die Lebern herausgeholt, um das Enzym zu isolieren, oder zumindest einen Extrakt daraus zu machen, womit man die transfer RNA mit radioaktiver Aminosäure beladen konnte, und wir haben sogar Prostata-Diphosphoresterase gemacht. Fig. 15 zeigt eine Seite aus dem ersten Jahresbericht des Instituts. Damit will ich nur zeigen, daß alles damals per Hand getippt wurde und höchst simpel in der Form war. Wenn man sich dagegen heutzutage Jahresberichte von Instituten ansieht, dann schämt man sich. Max hätte gesagt: "I don't want to have any of these". Max fand sie grauenhaft, Bilder von Labors und seinen Beschäftigten.

Hier bin ich an dem berühmten LKB Fraktionssammler (Fig. 16), der nie verläßlich funktionierte und immer mitten in der Nacht stehenblieb, bei dem also die wichtigsten Fraktionen immer danebenliefen — grauenhaft. Das ist Manfred Schweiger, ein Mitarbeiter, in

- 39 -

Arbeitsbericht für die Zeit von Oktober 1961 bis Dezember 1962

Der Arbeitsgruppe gehörten an:

Dr. H.G. Zachau
Dr. W. Frank (bis Sept. 1962)
Dr. W. Karau (seit Januar 1962)
Dr. D. Dütting (seit Mai 1962)
Dr. H. Feldmann (seit Sept. 1962)
Dr. R. Böhm (von Mai bis Okt. 1962)
Dipl.-Chem. F. Melchers (ab Nov. 1961)) als
M. Schweiger (bis März 1962)) Studenten
H. Heusinger, G. Patzelt u. S. Notz als techn.Assistentinnen
Frau P. Prüfert als Spül- und Laborhilfe

Die Arbeiten der Gruppe auf dem Nucleotid- und Nucleinsäuregebiet stellen eine Fortsetzung der vor dem Berichtszeitraum in München am Max-Planck-Institut für Biochemie begonnenen Arbeiten dar. Das Schwergewicht lag bei Arbeiten an der Fraktionierung und Struktur der löslichen RNA+) (A). Es wurden auch Versuche zur Isolierung hochmolekularer RNA und Vorversuche zur Fraktionierung dieser RNA ausgeführt (B). Außerdem wurden die Arbeiten an der Struktur und Synthese der Aminoacyladenosin-Endgruppe der Aminoacyl-RNA fortgeführt (C).

+) Abkürzungen: RNA = Ribonucleinsäure
DEAE = Diäthylaminoäthyl
AMP = Adenosinmonophosphat
GMP = Guanosinmonophosphat

Fig. 15. Annual Report of the Institute of Genetics, 1961/62.

einer typischen Pose nämlich schlafend (Fig. 17). Er studierte gleichzeitig Medizin und Biochemie und war die meiste Zeit am Schlafen. Das ist die berühmte Tee-Stunde unseres Labors (Fig. 18). Der „Doktor" (Zachau), wie wir ihn nannten, sitzt in der Mitte seiner

Fig. 16. Fritz Melchers at the LKB fraction collector.

technischen Assistentinnen. Man sieht auch Herrn Frank, Schweiger und mich selbst. Betriebsausflüge gab es damals auch, die waren immer sehr lustig. Zachau war ein ausgesprochen sozialer Mensch, der seine Seminar-Honorare in Betriebsausflüge investierte.

Jetzt komme ich ein wenig auf die Frage zu sprechen, was für einen Studenten wie mich so ungewöhnlich an diesem Institut war, abgesehen von einem weiteren wissenschaftlichen Horizont. Das erste war, daß Zachau mir eine Assistentenstelle, also eine Doktorandenstelle anbot, für sage und schreibe 450,- DM im Monat. Das war etwa das Doppelte des Wechsels, den ich von meinen Eltern bekam. Damit waren meine Eltern von finanziellen Nöten erlöst und ich vom Westdeutschen Rundfunk, für den ich als Kabelhilfe beim Fernsehen arbeitete, und wo ich Herrn Höfer und den Millowitsch abfotografiert habe. Das war alles Gott sei Dank vorbei und Zachau sagte dazu nur: "Safety enables", was ich ihm nie vergessen werde, denn das war wirklich wahr.

Das zweite war, daß wir das ehemalige „Häuschen" des Instituts beziehen durften (Fig. 19). Für sage und schreibe 15,- DM im Monat

Fig. 17. Manfred Schweiger.

mieteten wir ein Zimmer im „Häuschen" und das war natürlich eine Riesenverführung. Zum einen war man nahe am Institut bzw. am Labor und zum anderen gab im „Häuschen" phantastische Partys. Meine damaligen Mitbewohner Carlos David, Manfred Schweiger und Ekkehard Kölsch wissen wovon ich rede, und Charly Steinberg hat seine Frau dort gefunden.

Was ich sagen will ist, daß Köln in einem Maße lustig war, wie es für einen Doktoranden eigentlich selten war. Es gab die sogenannten ERÜPAS (Entrümpelungspartys), zu denen jede Teilnehmerin und

Fig. 18. Teatime at the Zachau lab.

jeder Teilnehmer ein Stück mitbringen durfte, das dann versteigert wurde. Sie werden kaum glauben, aber das wurden sehr lustige Feste und sehr alkoholische Feste, bei denen unglaublich große Mengen Geld zusammenkamen für den größten Schrott, den man sich vorstellen konnte. Die ERÜPAS haben eine Art von Tradition übernommen. Später als wir in Berlin waren, gab es solche ERÜPAS bei Thomas Trautner am Max-Planck-Institut für Molekulare Genetik, weiter hier in Köln und dann ab und zu einmal in Freiburg. Es gab den berühmten Löwen, der immer geklaut wurde, sozusagen die Trophäe dieses Festes. Zu guter letzt ist der Löwe leider gestorben, weil ihn irgendjemand hingeschmissen hat. Auf den nächsten Bildern sehen Sie Thomas und Gerda Trautner in jungen Jahren (Fig. 20). Und das ist Carsten Bresch, der angeblich nirgendwo auftritt: Erkennen Sie ihn? (Fig. 21)

Nun, woher kam das Feiern? Max wollte diese Feste. Er sagte, er wolle keine Party im Institut ohne ein ordentliches Programm und wir sollen uns etwas einfallen lassen. Diese Tradition der Institutspartys mit selbstgestalteten Programmen hat sich fortgesetzt zumindest in Basel, Maria Leptin kann ein Lied davon singen. Hier ist Joseph Straub in einer etwas verfänglicheren Lage mit der Botanisiertrommel, also auch Joseph Straub war ein Fan von solchen Festen. (Fig. 22) Sie können sich vorstellen, wie gut wir es gehabt haben.

Fig. 19. The provisional little lab and living house "Häuschen" for the geneticists near the Botanical Institute.

Fig. 20. Gerda and Thomas A. Trautner.

Ich will am Ende sagen, daß diese Zeit hier in Köln für mich die Basis für mein späteres Leben in der Wissenschaft war, was ich am Anfang gar nicht so begriffen habe. Gerda und Thomas Trautner wurden große Freunde. Wir haben sie in Berkeley wieder gefunden, und er hat mir nach meiner Rückkehr aus den USA meinen ersten Job in Deutschland angeboten, am Max-Planck-Institut für Molekulare Genetik in Berlin. Es war schon ein Network vorhanden.

Als ich dann ans Basler Institut für Immunologie ging, hat mir Rainer Hertel zusammen mit Carsten Bresch in Freiburg die Habilitation möglich gemacht. In Freiburg fand ich auch meinen ersten Doktoranden, nämlich Georges Köhler [Nobelpreis 1984]. Max Delbrück hat mir eigentlich die ganze Karriere aufgebaut, weil er mich mit Ed Lennox zusammengebracht hat. Durch ihn bin ich am Salk Institute gelandet und anschließend auch bei Niels Jerne. Im wesentlichen trieben mich die Kontakte, die ich durch Max aufgebaut habe. Und wenn ich so ehrlich sein will, dann ist auch Dieter Dütting an meinem Glück schuld, denn er hatte in seiner Kölner Zeit eine technische Assistentin, Gisela Schultz, die zusammen mit der Dame wohnte, die

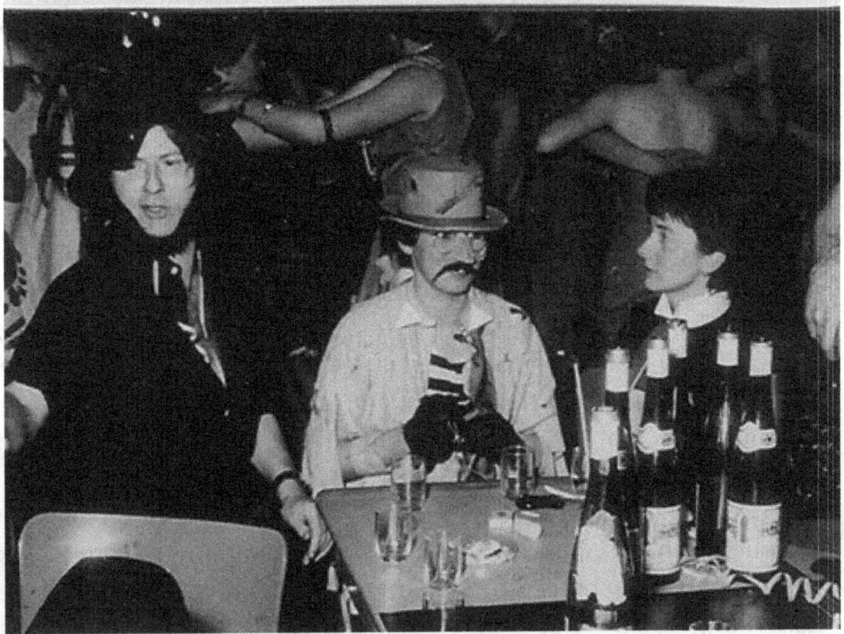

Fig. 21. Carsten Bresch (*left*).

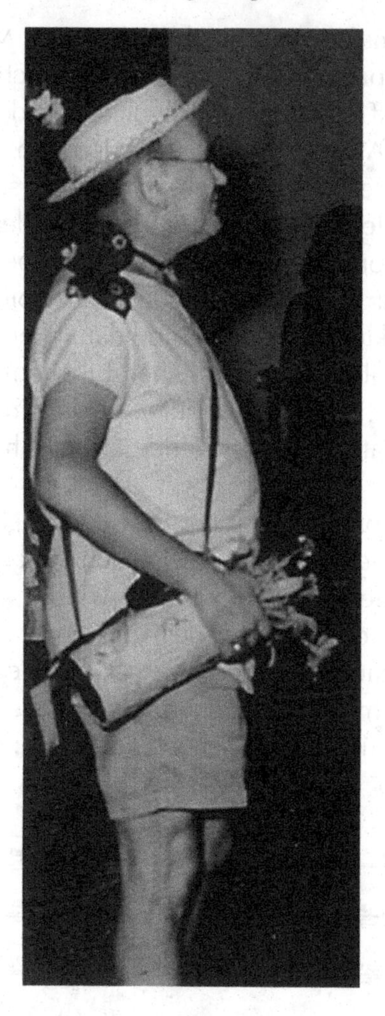

Fig. 22. Joseph Straub with the botanist's container (Botanisiertrommel).

jetzt seit über 40 Jahren meine Frau ist. In Köln ist einiges geknüpft worden. Wenn ich auf mein Leben zurückschaue, ist eigentlich alles nur diese Kölner Verbindung gewesen. So haben wir viele Traditionen, die hier in Köln und vorher vielleicht in Pasadena begonnen wurden, in Basel weiter geführt.

Ich komme nun zum Ende und sage, daß das, was, mich eigentlich in die Wissenschaft getrieben hat, eine ganz simple Sache war. Wir arbeiteten an transfer RNA und hatten das Modell von Giulio Cantoni, ein Hairpin-Modell, das auf der geschlossenen Seite ein Anti-Codon und auf der anderen Seite mit den beiden offenen Enden die Aminosäure hatte. Als wir schon viele Teilsequenzen hatten, habe ich auf einmal festgestellt: Das kann gar nicht so sein. Das kann kein Hairpin sein. Das letzte Bild zeigt die Struktur der transfer RNA, wie sie in meiner Doktorarbeit steht. Sie ist falsch, aber kein Hairpin, sondern ein Kleeblatt (Fig. 23). Zachau sagte sofort, solange er die genaue, totale Struktur nicht kennt (und damit kein Molekulargewicht angeben kann), gibt ein ordentlicher Biochemiker nichts heraus. Diese simple Einsicht, daß transfer RNA kein Hairpin ist, sondern eine Kleeblattstruktur hat, diese aufregende Einsicht hat mich nicht mehr ruhen lassen.

Nach meiner Promotion mit der unvollständigen Struktur ging ich ans Salk Institute. Bevor ich ging, zeigte ich Max meine Doktorarbeit. Er fing an, sie zu lesen. Dann kam ich wieder zu ihm und meinte: „Zachau hat gesagt, das ist alles geheim, das darfst Du noch nicht weiter erzählen". Max machte meine offen vor ihm liegende Doktorarbeit wieder zu, sein Name stand schon in seiner Kopie der Arbeit. Er schob mir das Exemplar über den Tisch zu und sagte: "Whatever you tell

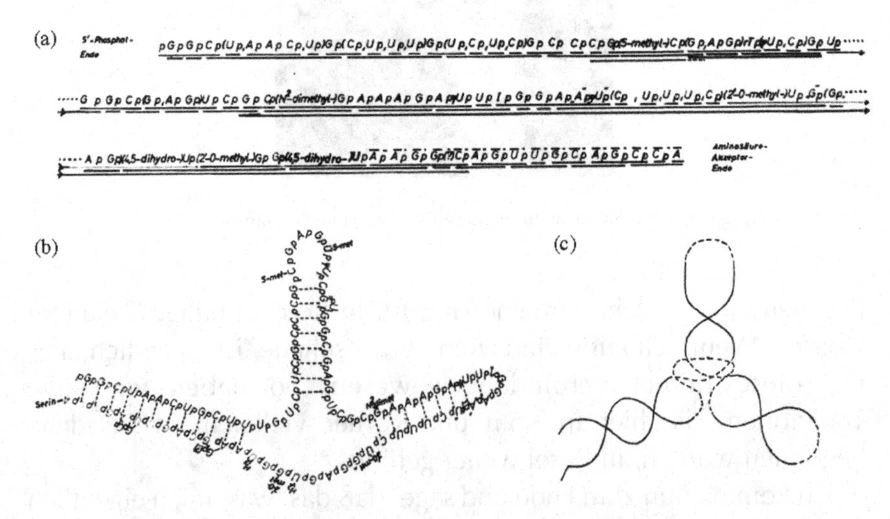

Fig. 23. Sequence of transfer RNA as shown in Fritz Melchers' dissertation.

me, consider it published". Das habe ich mir gemerkt. Als später Roche Forscher in Basel am Institut uns „geheime Forschungsergebnisse" aus ihrem Institut erzählen wollten, sagte ich zu Ihnen: "Whatever you tell me—consider it published."

Als ich nach La Jolla kam, sind die Leute über mich hergefallen und haben gesagt: „Du hast doch bei dem Zachau gearbeitet." Die Leute waren Monod, Crick und Orgel. Sie sagten zu mir: „Wir sind dabei, transfer RNA Strukturen zu basteln, Holley hat gerade die erste Struktur veröffentlicht. Wie ist es denn mit Eurer, gibt es da vielleicht eine generelle Struktur?" Da habe ich gesagt: „Ja ja natürlich und obwohl es geheim war, habe ich es ihnen erzählt. Das war nicht so gut, denn Zachau fand es nicht gut".

Drei Wochen später fuhr ich auf den Antibody Workshop. Dort erzählte mir Norbert Hilschmann am Abend vorher, was er gefunden hatte, nämlich die variablen und konstanten Regionen der Antikörpermoleküle. Dann sagte er: „Ich weiß gar nicht, ob ich das erzählen soll und ich hab es auch noch nicht veröffentlicht". Da habe ich ihm meine Geschichte der geheimen Struktur der transfer RNA erzählt. Am nächsten Tag hat er es nicht für möglich gehalten, seine Entdeckung ganz zu erzählen, obwohl alle Größen der Molekularbiologie, unter anderem Crick, Watson und Delbrück, anwesend waren. Nachdem er die kompletten Aminosäuresequenzen von drei leichten Ketten von Antikörpern kurz gezeigt hatte, aus denen klar wurde, daß die Ketten aus variablen (am N-terminalen Ende) und konstanten Regionen (am C-terminalen Ende) aufgebaut sind, weigerte er sich, die genauen Strukturen in der Diskussion noch einmal zu zeigen. Jeder wollte wissen, wie die von ihm postulierte Rekombination von vielen V-Segmenten auf ein C-Segment erreicht werden könnte. Da er diese Information nicht wieder gezeigt hat, haben ihn alle Größen der Molekularbiologie beschimpft und ihm gedroht, er würde nie wieder auf ein solches Meeting eingeladen werden. Hilschmann hat an dem Tag sein ganzes Glück verloren. So, was soll ich sagen. Ich habe hier in Köln etwas gelernt, was vielleicht eine Hypothek ist, aber ich bin glücklich geworden mit dieser Hypothek.

13

The Institute's Impact on Neighbouring Disciplines

Georg Michaelis

I think I was the third chemist, after Fritz Melchers and Horst Feldmann, who switched from organic chemistry in Cologne to the Institute for Genetics. I studied chemistry at the University of Cologne from 1958 to 1964 and completed my Diplom. Initially, I wanted to study biology and not chemistry, but my parents, both biologists, were convinced of the future role of chemistry in biology and they were sceptical about the prospects for a tenured position in biology. So they offered to pay for my studies in chemistry. At that time the chemical industry played an important role in the German economy. Sadly, this leading role has since declined considerably — and the decline continues today. After obtaining the Diplom in chemistry I wanted to change to botany, the field I finally ended up in. I participated in the full-time practical course in botany, but after this course I decided to look for a topic for my doctoral thesis at the Institute for Genetics.

The reason for this change was the hierarchical atmosphere in the Institutes of Chemistry and Botany, which contrasted with the climate at the Institute for Genetics. I was looking for maximum freedom and developed a certain aversion against hierarchical attitudes. For example, I had worked as a *Werkstudent* in the Bayer factory to earn some money and had to buy glass equipment for the laboratory courses, which was rather cheap there. When I wore a white coat, most people in the factory were polite and helpful, asking what they could do for "*Herr Doktor*". The grey coat worn by technicians evoked quite a different response and one was asked "*Was willst Du heute?*" ("What do you want today?"). The personality of Professor Otto Bayer, who gave lectures in the Institute of Chemistry, left an authoritarian impression on me.

For these reasons I did not want to work in a chemical plant like the Bayer factory. I remember a young Privatdozent as a guest speaker in the seminars of the Institutes of Chemistry who mentioned that he had obtained permission from an established professor in the field to perform a particular experiment. For a young scientist like me, the prospect of having to ask for permission to perform experiments was not very appealing. I also remember an occasion in the Institute of Botany when an assistant was publicly corrected by the professor regarding his remarks of important scientific problems in botany.

In retrospect, it is interesting to compare the styles of scientific publications from the Institutes of Chemistry and Botany. In chemistry, the sequence of authors was determined by seniority: the professor was the first author, followed by the habilitated co-worker, the graduate students and finally by the students working for their Diplom. As students, we considered this procedure quite normal and did not criticise it; we were happy to be listed as co-authors. In the Institute of Botany, publications were handled quite differently, influenced by distinguished personalities like Professor Otto Renner. Professors like Georg Melchers, Joseph Straub in Cologne, and Wilfried Stubbe in Düsseldorf generally allowed their Ph.D. students to publish alone.

I once happened to remark that a particular professor hadn't published very much, but I was corrected by a colleague, who explained this style of publication: no co-authorship of Ph.D. work. Even today I'm fascinated by these differences in attitudes to publications, and I wonder how these publications were rated in the DFG referee system. As an example, let me cite the publication: Birkofer L., Kaiser C., Kosmol H., Romussi G., Donike M., Michaelis G. (1966), D-Glucose- und L-Rhamnoseester der p-Cumar- und Ferulasäure. *Liebigs Ann.*

Chem. 699: 223–231. This paper was the result of a cooperation between the Max Planck Institute in Cologne-Vogelsang and the Institute of Organic Chemistry. At the Max Planck Institute, Drs. Dieter Hess and Wilhelm Seyffert studied the genetics of the anthocyanin biosynthesis. The structures of the glycosides and their intermediates were determined at the Institute of Organic Chemistry. A second botanical publication, by Horst Binding, who later became a professor at the University of Kiel, may serve as a counter-example: Binding H. (1966), Regeneration und Verschmelzung nackter Laubmoosprotoplasten. *Zeitschrift für Pflanzenphysiologie 55*: 305–321. Binding himself, then a student of Professor Joseph Straub, is the sole author of this paper. These cases illustrate the attitudes to hierarchy in chemistry and botany in Cologne. The Institute of Genetics was quite different: a new scientific field was represented not by one professor but by several young professors who tried to transfer the department system founded in the United States to this University and to this new Institute lacking any tradition.

The *Ring-Vorlesung* in genetics was unique and very attractive for students. The important feature of this lecture was that all the genetics professors and their co-workers attended. Therefore, the lecturer had to be well prepared and very often the lecture was interrupted by questions, which sometimes initiated a general discussion. This was interesting for students because controversial views or topics were presented. I remember a lecture on electron microscopy in cell biology, given by a zoologist from Cologne, Professor Köcke. A lively discussion ensued, although the geneticists were not really competent to distinguish between artefacts and meaningful biological structures in the electron microscopic photographs. Indeed, in my own field of research, the mitochondrial genome, this is still a controversial issue. Mitochondrial genetic maps are circular, and photographs of circular mitochondrial DNA molecules have been published for fungi and higher plants. However, only recently, Professor Arnold Bendich (Seattle) studied mitochondrial DNA of higher plants and discovered large linear molecules.

Besides the non-hierarchical atmosphere and the "Ring-Vorlesung", the various research topics investigated by the four or five professors of genetics were a third reason why the Institute of Genetics was attractive but also difficult for students. The different special topics were one reason why the seminars in genetics were sometimes difficult to understand. I remember an encouraging remark made by Professor Zachau, a biochemist at the Institute, explaining that it was not necessary to understand everything in the seminars. He himself demonstrated that outstanding research in molecular biology was possible without a detailed understanding of special topics in genetics.

Finally, the enthusiasm for science and research in the Institute of Genetics should be mentioned. Very often I worked at night in the Institute. Standing outside the building around midnight, the contrast with a dark Zoological Institute on the left and a dark Botanical Institute on the right was striking. The illuminated rooms on the different floors of the Institute of Genetics vividly suggested exciting activities and experiments going on in its laboratories.

I have mentioned four reasons why I did not stay in chemistry or botany and why I considered the Institute of Genetics in Cologne to be very attractive for students at that time. I did the work for my doctoral thesis on the sequential appearance of the enzymes of the galactose operon in the laboratory of Professor Peter Starlinger from 1964 to 1968. I have some very happy memories of this period. The annual course on bacterial genetics attracted scientists from all over Germany, and friendly relationships developed from these meetings. In addition to the bacterial genetics course, there was the Spring Meeting on current research topics. Many invited speakers presented interesting seminars, and the seminars or lectures given by Max Delbrück were also impressive. I once complained that having regular seminars three days per week did not leave enough time for experimental work. But Peter Starlinger replied that two days for experiments should be sufficient, provided the experiments were carefully planned and executed. This research atmosphere was not always easy for a chemist like me. (Chemists were sometimes accused of being good experimenters but not very incisive thinkers!). However, the many visiting scientists from all over the world contributed to wide-ranging discussions and provided much stimulation for research.

Finally I should mention a successful initiative that had its origins in the Genetics Institute. The initiative was provoked by questionable refusals of manuscripts from German sources by American and English journals like the *Journal of Molecular Biology*. Several prominent German scientists therefore decided to submit their best manuscripts first to the oldest journal of genetics, *MGG* (*Molecular and General Genetics/Genomics*). Given the ever increasing dominance of a small number of "flagship" journals, and the perceived importance of impact factors, I should say that *MGG* perhaps needs such an initiative again. My last grant application to the DFG was criticised by a referee, apparently a biochemist, who advised that I should do less genetics and more biochemistry, and asked why I published my results in a journal like MGG. If *MGG* fails to survive, I'm convinced that we will lose an important tool in science, increasing our dependence on American journals and referees.

Later in my career I worked in institutes of biochemistry, molecular genetics, physiological chemistry, molecular biology and botany, and was sometimes confronted with prejudices against genetics. One instance of this happened at the University of Bielefeld during the search for the first professor of genetics to be appointed there. The main research interest in the biology faculty was ethology, and there was a consensus that this focus should be maintained. Most professors were wary of a strong genetics institute like that in Cologne (some of the botanists at Bielefeld had worked for some time in the Botanical Institute in Cologne and were quite familiar with the situation there). One candidate explicitly stated that the curriculum for the Diplom in biology at Bielefeld should be changed and modernised. He was of course not selected (he later obtained a position at the University of Bonn). The person who was finally selected made a very unassuming impression at the time. Unfortunately for the primacy of ethology at Bielefeld — but luckily for the biology faculty — the new geneticist established a very strong and successful genetics department there (something the search committee obviously did not expect of him!).

In 1969 I left Cologne, a city with a tram system but no metro. With the exception of Sankt Aposteln, all the Romanesque churches were still closed because of reconstruction and repair work, necessitated by the damage caused by the second world war. In the meantime, the building that housed the old Institute of Chemistry — in which Professor Kurt Alder carried out the work that earned him a Nobel Prize — has been demolished. The Institute of Genetics has moved to new laboratory facilities. The Botany Institute will follow soon.

Looking back, it has been a long scientific journey from Cologne, via the University of California at Davis, the CNRS in Gif-sur-Yvette (France), Würzburg and Bielefeld, to Düsseldorf. When I compare these different places, I must conclude that the old Institute of Genetics here in Cologne was one of the best places, perhaps the best of all, in which I had the privilege to do scientific work.

14

Life with Bacteriophages

Thomas A. Trautner

With my remarks I am covering an important period of my life, spent in Cologne between 1957, when I obtained my doctoral degree, and 1964, when I accepted an assistant professorship at the University of California in Berkeley. Within this period I spent one and one half years as a post-doc in Arthur Kornberg's lab in Stanford. Also during this period, I had my first doctoral student, Elke Rottländer, who is in the audience. On the private side, I got married in 1959, and our daughter Tamara was born in 1963. At the time one could live comfortably and very relaxed at the Institute of Genetics in Cologne without grave concern about one's own future — without all the rules and constraints which make scientific life much more difficult these days.

Until 1960 we worked under the heading of the "Abteilung Mikrobiologie des Botanischen Instituts". It is worth remembering that at this time only one chair for non-medical microbiology was in existence in Germany. During this early period, the main thrust of our group was on bacteriophage genetics. As mentioned earlier this morning by Carsten Bresch, one of the questions of the time concerned the "mating theory" put forward by Visconti and Delbrück to explain some poorly understood results of phage crosses. Its general validity came under serious doubt following work by Bresch and Starlinger. For our work we used the *E. coli* bacteriophage T1. This followed the previous discovery of Bresch that certain colour additives to agar plates containing high glucose would stain bacteriophage plaques. The type of stain was genetically determined and, with colour mutants and other available mutants of phages, permitted the scoring of all eight types of progeny from a three factor cross. This technique was extremely helpful in our analysis of the effects of the multiplicity of phage infection on recombination frequency or experiments dealing with phenomena related to phenotypic mixing, observed with serologically distinguishable T1 phages.

Our approach to phage genetics did not require much technology; we mostly followed the credo of "*viele (Petri-)Platten — scharfes Nachdenken — ausführliche Diskussionen*" with our colleagues Charley Steinberg, Gus Doerman, Rudi Hausmann, Rainer Hertel and others. Discussions of our own experimental work were complemented by an intensive seminar programme both with outside speakers and within the group. I remember very well one of the most fascinating seminars, which Peter Starlinger organised. It dealt with immunology; in particular, with the contrasting theories (instructive vs. selective mechanisms) put forward by Linus Pauling and Joshua Lederberg.

My research interest changed after I returned from Stanford to Cologne in 1962. I had been exposed to the biochemistry of DNA. Under the influence of Dave Hogness and Dale Kaiser, who had developed a system of infection of *E. coli* by purified lambda DNA, and after learning about genetic transformation of B. subtilis in the neighbouring department of genetics by Joshua Lederberg, I used competent cells of B. subtilis as recipients for infection with DNA from a new bacteriophage, which we had isolated from the flower beds of the botany department next door. We called this process "transfection" — a recombinant word between *trans*formation and in*fection*. The transfection experiments, which we performed with phage DNA, turned out to be interesting in the sense that the dose response between infective centres produced and DNA concentration was exponential and not linear as in standard phage infection. We attributed this to the mechanism of DNA processing during uptake: double stranded DNA would obligatorily be converted to a single stranded form. Only when a complementary strand from another transfecting molecule was available would the strands anneal intracellularly and produce a double-stranded DNA molecule, representing a template for DNA replication. From the observation that non-phage or B. subtilis DNA competitively inhibited the efficiency of transfection, we concluded that DNA uptake into competent B. subtilis cells was not specific.

This observation led to cooperation with Pamela Abel in the department, who had described a subviral entity of vaccinia virus, whose infectivity was DNAse sensitive. Experiments, in which we used such material for the transfection of B. subtilis cells apparently led to the generation of intact vaccinia virus. These results were in our hands not reproducible and were subsequently revoked.

The experiments with phage transfection, however, were expanded to involve plasmid transformation, and led to a unified picture of DNA uptake into competent cells of *B. subtilis.* Such experiments included investigations on the effects of restriction and modification on DNA processing, which eventually opened a further area of research performed in Berlin and focussed on DNA methylation. This research also involved bacteriophages, with the discovery that a number of B. subtilis phages carried genes for DNA methyltransferases. These genes were very peculiar in the sense that the enzymes they encoded were multispecific, which opened the way to define the domain structure of target-recognising domains of such enzymes.

I was happy to accept the invitation of the organisers to this meeting. It is important to conserve views of eye-witnesses of scientific developments. In particular, I was happy to see that your announcement of the meeting showed a facsimile of the letter of Joseph Straub to the Deutsche Forschungsgemeinschaft pointing out the need for the foundation of the Cologne Institute of Genetics. Straub, with his vision and his energy to realise his plan, was the *Urvater* to whom we owe the presence of this excellent institute.

IV. Views from Outside

15

View from (Cologne) Physics

Bernhard Mühlschlegel

I joined the faculty in Cologne as a young theoretical physicist in the spring of 1962. A few days after my arrival, I was invited by Max Delbrück for dinner in his house in Bachemer Strasse, and so began a lifelong friendship with Max and Manny. I took part in the great opening ceremony of the Genetics Institute in the summer of 1962. As a natural consequence, a nice contact emerged between the Physics and Genetics Institutes, which remained active also in later years after the Delbrücks had gone back to Pasadena.

The regular faculty meetings brought together all of the science professors, and this was a rather conservative group of people. The activities in the new field of genetics were not always welcomed by the rest of the faculty. Some older colleagues did not like Delbrück's *American style*; some traditional biologists were concerned about the growing attraction of genetics in comparison with their own research. I remember a few *voting battles* on important issues (also after Delbrück left), and I feel good since genetics won due to a clear support by physics in these votes. However, occasionally there was also voting without the desired success. The ritual of *Habilitation* also served as a power play between different groups in the faculty.

How about the chances for a true scientific cooperation between the Genetics and the Theoretical Physics Institute in Cologne? There were always common discussions when Delbrück was here till 1963, and also during his frequent visits later. There were also social events for both institutes. For instance, in the summer of 1964 a swimming competition was organised in one of the lakes outside of Cologne. Gold and silver for two theorists, bronze for a geneticist!

As we all know, Delbrück turned from physics to biology in the early 1940s. Perhaps he hoped that he would be an example for others, not to mention the words of heroes of the past like Bohr and Schrödinger. Actually, there are many people who started as physicists and later, but still at a young age, switched to biology. Among them, as I remember, Carsten Bresch, Werner Reichardt and Walter Gilbert. In Pasadena, Delbrück had post-docs and graduate students from physics doing *Phycomyces* research. One of them by the name of Ernst Peter Fischer had a very good diploma in theoretical physics from Cologne. After a Ph.D. in *Phycomyces* studies, he became a Delbrück biographer and is now a popular science writer in Germany. In Cologne, Delbrück and I tried to persuade Klaus Dransfeld from Berkeley to accept a chair here. He was an excellent experimentalist, very much interested in molecular biology, and he appeared to us the ideal candidate for building a bridge between genetics and physics. But, unfortunately for us, he remained in condensed matter physics and went to Munich, Grenoble and finally to Konstanz.

Coming back to the question raised above, one must admit that a true scientific cooperation of physics with the Institute of Genetics in Cologne could not be established in the 1960s and the following decades. One might regret this, but one has to realise that both fields flourished immensely on their own. In condensed matter physics, for instance, many new phenomena of electronic properties in solids showed up, and research done in Cologne gained an international reputation. The series of *Sonderforschungsbereiche* in genetics and physics have demonstrated the quality of many different activities in our sciences over the years.

Time goes on; the old actors are *in Rente* or dead. The number of researchers has increased remarkably, as has the number of scientific topics treated. And the power of experimental possibilities and theoretical and numerical methods are today at a level which one could barely dream of forty years ago. It is then very encouraging to observe that Michael Lässig and Diethard Tautz have started a new cooperation between physics and genetics. This is a first step, and I wish them lasting success for their endeavour.

[Added in November 2005] It is even more encouraging to learn that the cooperation between physics and genetics in Cologne was recently put on a broader basis by the successful foundation of the new *Sonderforschungsbereich* 680: *Molecular Basis of Evolutionary Innovations.* I am sure also Delbrück would have strongly supported these new developments.

16

In the Smog of Genetics: Biochemistry in Cologne — My Version of History

Lothar Jaenicke

Medical chemist in Cologne

Coming to Cologne from Theodor Bücher's Marburg via Feodor Lynen's Munich, I first had an appointment at the Medical Faculty as newly established *Extraordinarius* (Associate Professor) of Physiological Chemistry. There was a well known, and just enlarged, Physiological Chemistry Institute in a fine new building, authoritatively ruled by the strong-headed, combative Ernst Klenk of international neuraminic (sialic) acid and complex brain-lipids fame. He was a natural products chemist, dyed in the wool, well organised and with wide interest in structures and diseases (less in metabolism) of normal, deviated and demyelinated grey and white matter. He accepted me as if he had come to this junior colleague somewhat like the proverbial virgin to her (Jesus?) child and behaved accordingly. Sensing it was tenderly amusing.

Up to that date, there was not felt a need for biochemistry as we know it now at the Cologne Natural Science Faculty, which recently branched off the Philosophical Faculty. It was still too new to them. It was only fifty years ago that it began to evolve in Dahlem with its roots in Otto Warburg's Kaiser Wilhelm Institute of Cell Biology and Otto Meyerhof's Physiology Section at the Kaiser Wilhelm Institute of Biology, and later at the Kaiser Wilhelm Institute of Medical Research. Soon, however, most of its ambitious proponents were forced to emigrate. They transferred their knowledge overseas, much to the detriment of Germany. "Hitler's Gifts" from the Warburg, Meyerhof and Neuberg entourage became proverbial. Theodor Bücher, later in Marburg, was Warburg's most powerful local descendant. Biochemistry hibernated. It was nourished in Heinrich Wieland's genetically fertilised soil and flowered after Liberation in Feodor Lynen's Munich University and Max Planck (Karlstrassen)-Institutes, mostly working with Bavarian brewer's yeast on acetate metabolism.

Kölsch-favourising Cologne at that post-war time, had a particular "Institute of Fermentation Sciences", an extricate offshoot of Max II Delbrück's uncle Max I's Berlin "Seestrassen-Institut" (Institut für Gärungforschung). The history of its founding and its spirits had a special local flavour. Its director, Hermann Fink, training brewing masters and also working on dietary problems in rats (of which posthumously came out selenium as essential micronutrient!) died unexpectedly in 1962, and when looking for a successor, eyes fell upon me; eyes, as far as I know, among others of Joseph Straub, the botanist who was the moving power behind reforming Cologne Biology and of Max Delbrück who, as is well known, as a physicist ("working by clean thinking on clean compounds", in contrast to the chemists who might have pure stuff yet impure reasoning and, even worse, the physical chemists who had neither - there were no biochemists to slander at that time) did not estimate this science highly, however, *faute de mieux*, was *d'accord*.

The chemists were less enthusiastic for other reasons but finally, albeit hesitantly, accepted biochemistry as a fifth wheel on their carriage. And influential Ernst Klenk, then *Rector Magnificus*, was happy to get rid of an incubus in the bed, well made to his size.

Denizen-biochemist, science faculty

I was also happy to accept the parting, the honour and the challenge to found, in 1963, the first Institute of Biochemistry north of the river

Main, the "*Weisswurst-Äquator*". (*Weisswurst* is special Bavarian sausage made from (naturally BSE-free!) pig brain, among other ingredients, to be relished before noon with sweet mustard and appropriate beers. In *Hillige Cologne* it is replaced by less bland "Halve Hahn", the recipe of which will not be given away here).

The "IBCh/UKln" remained in its provisional founding status for almost 40 years, in a romantic medieval setting at the end-corner of the Old Southern Town, called the "Bottmühle" (from a museally conserved post windmill-without-wings on a platform of the cyclopic medieval town wall). It was very private and far away from all faculty happenings, but growing outriggers in different places due to its increasingly heavy teaching load (mainly for biology students) and growing team. Before the present institute was built and occupied in the mid-1990s, there were ultimately three part-institutes; one with the laboratories for the students' classes, but no auditorium, so that lectures had to be held at the mathematical, physical or geo-sciences premises, moving around like Charlemagne from court to court, yet by bike instead of ox-cart. In its place, a hand-cart or sturdy bike-trailer was used to move through town equipment required for the practical courses.

As it happens not only in faculties, there were frictions over birth rights and legitimacy, alimentation, funds, and needs, even before "excellence" (i.e. modern "von" Universität) was invented. The new-born baby grew up under strained standard conditions, but it grew. It just *had* to grow under these proverbially sturdifying circumstances.

Max's aura

An old saying goes: "The prophet does not count in his own father-land". The fame of the "Delbrück-Institute" was world-wide. But in Cologne its fame was ambivalent and diffuse, not reaching much beyond the Science Faculty. In the Medical Faculty, to which I belonged to begin with, knowledge of this jewel in the University's pious crown was only dim. After finally agreeing to have an Institute of Biochem-istry in 1963, faculty, in voting, was, as usual, divided into the self-interested, the disinterested and, let us say, the Humboldtian-emotionals.

In genetics-prone agenda, there were the Delbrück-collegial, yet materially uninvolved physicists and mathematicians (about half of the total), the constant laager of strictly fenced-in chemists, always on the lookout for invaders, and the growing number of self-centred bio-scientists still conserved in *quasi*-Linnean boxes. And where was wobbling biochemistry? The eventual outcome of voting on both contentious and undisputed matters between these groups could be foreseen.

I knew the futuristic rumours on the GENE and the extrapolated scientific problems, a little bit of steam-engine genetics, and virology and was eager to meet and contact the new colleagues. The best means was to understand the environment by having a look at the promising crew by approaching them personally and attending their advertised, yet limited (to about twenty participants), spring course that year (1964) in phage genetics. It was organised on the blueprint of the famous Cold Spring Harbor pattern by Carsten Bresch, Thomas Trautner, Rainer Hertel, Michael Reusse and Wolfgang Michalke under the unmistakable aegis, of course, of "MAX", and with the oral aid of the rest of the genetics team: Peter Starlinger, Walter Harm, Hubert Kneser and Hans-Georg Zachau. I was duly impressed. This was a thrilling event as well as a challenge in any aspect: contents, performance, intensity, intellectuality, fervour, and whatever else.

It remained a lasting experience. I tried to meet it at a more pedestrian, situationally limited level in my parochial biochemistry courses for which I had written a curriculum, a lecture script and a methods book, long in use. Laudable zeal to compensate frust by facts. Jane Austen would have titled: Drudge and disillusion.

The then new Institute of Genetics was by no means a palace. It was economically built, even unusually plain, at a time when other institutes were spaciously planned with big showy entrance halls. An installation many were envious of was a room with the necessary facilities for invitees, that was often occupied by the many scientists from abroad that gave the institute its characteristic flair — in contrast to such guest-rooms in other institutes that were mostly used as the private quarters of the boss, planned for retirement. Whereas in other new buildings you found long floors studded with labs, with labels on doors that were generally closed, but without function inside when you dared opening them, here the working rooms had no doors and were amply equipped. And most equipment was accessible publicly as in the United States; science was discussed in competitive tug-of-war cooperativity with graded arrogance. When I shortly thereafter organised for a period a regular common genetics/biochemistry seminar, I was condescendingly commended for the fact that I, surprisingly, understood some actual immunology and developing molecular biology, despite being only a chemist with chemical leanings on trite enzymes and reaction mechanisms.

The mostly ups of genetics

The "critical mass", not yet fully completed according to plan, energised the reactor already. I, a kind, though at times suspect, yes-man and bad strategist, helped to fill the gaps. Thus, we collected *ad maiorem Maxi gloriam* a harvest of prospective eminence and efficiency from fresh talent here and overseas. At a cost.

It "came as it had to come": with limited funds (and when are funds not limited?) growth, evolution and development of one side could only be sustained at the cost of another, namely the weaker. That is applied — not only academic — Darwinism; well understood and handled in biology circles as "New Synthesis", depending on populations, not single individuals. One-man's biochemistry could not dwell and *evo-devo* in a niche; it had to grow out of its limited base, not so much for reason of standing than of need: teaching, research and crowding required more space than available at the Bottmühle. Where to take it without stealing? The chemistry colleagues were happy in their spacious post-war-modern glass/concrete ersatz for the old bogus-style brick/neo-gothic Augusta Hospital. The plans to integrate biochemistry were dropped in the interregnum, and the administration felt powerless to squeeze out some square meters for a hanging garden between the wings of the '**F**'-shaped architecture of the chemistry complex to make it a '**P**' for somebody like me. And, I must confess, I was afraid to become a permanent thorn in the solid chemical flesh.

The day of a biochemist

Lecture at 8 am with awakening early morning exercises in chalk and talk. Afterwards handling *manu proprio* (in a lab coat, professional, yet indeed unusual for a veritable professor) the practical courses, teaching the beginners the Little Art of biochemical doing, calculating and reasoning till noon, when I had to cycle down-town to the Bottmühle to do the rest of my duties; to keep the deck clear and things going. This timetable did not leave time for mensa-lunching and lobbying, nor even to stop at home with my family. Much to my loss and handicap, as it soon turned out.

There were very concrete plans to expand genetics to its capacity despite some misgivings of the biologists in neighbouring institutes. I heard of all of it mostly too late, when things had already been arranged and brought before the Faculty. Finally, in all the voting on the procedure, genetics won over biochemistry by a solid and loyal five to one, which remained unchanged despite the recently, but evenly appointed C3-professors. The "silent majority" of sharing onlookers, who mainly followed rumours rather than reason, was no help. Succour, if at all, came from the self-critical physico-chemist, some amiable botanists, disinterested ecologists and emotional souls who were at best expediency seeking colleagues, not logistically cooperating chess-mates like the genetic phalanx combined in their cause.

Tactics won over strategy. One time-marking compromise followed the other for many years. One of the hitting arguments against the need of biochemistry for biology in general, and genetics in particular, I still have in my ears.

The "biochemistry genetics needs..."

It was coming from the speaker of the involved institute: "The biochemistry we need, we do make ourselves" — no comment and curtain down. Biochemists remained denizens here and there.

True — for economic reasons, I left the baptismal holy water of biochemical genetics to the initiated. We had not more than two years for the basic curriculum in all of biology, and biochemistry was a must for all students.

It was only by brachial pressure from the students, that finally money was allotted for some new equipment and, much later, quite a sum for a temporary abode for biochemistry teaching. It was planned for ten years, and lasted three times that long, surviving me and all involved. The money definitely did not cut into the growth of the rest of the faculty. The country was still in a phase of pressure, under the spell of Pedagogy Professor Georg Picht's *Bildungsnotstand* (educational emergency). Today we would compare it to *PISA* on a higher level. And "excellence", with its expedient arena of impact fighters, was far away in the United States.

Spreading the gospel — SFB 74

But money for research did not lie on the streets. It had to be extracted from the German Research Council by either individual or combined effort. For that purpose, a new instrument was the "*Sonderforschungsbereich*" (SFB, Special Funding Bag). It needed, beyond a certain quality, also a certain critical mass — on the side of the reviewed as well as the reviewers. (I know it from both perspectives.) On planning the SFB 74 (with the all-embracing title "Molecular Biology of the Cell") the far-seeing and well-advised strategies looked for a flank-guard. I joined first with a topic on methyl group transfer, without knowing how essential this would become much later in virology and epigenetics. In a second round my topic was the development-inducing pheromone of *Volvox*, a protein. I was not alert enough to follow the early collegial advice to go at it by molecular genetics. And when we did it, clumsily self-trained much later, it was too late. Retirement was at the door. In one case a crucial line of research was stopped by profile-seeking embryo protectors. Unjustly, when the toddler already walked, the emeritus was rehabilitated: he had violated neither God-made nor man-made rules. As so much and so often, he took it with gallows humour.

For me the most interesting events were the SFB colloquia and the lectures — very early on in English, of course, even if there was no native speaker. All this was on such a high scientific level that I frequently had difficulty not to get lost in the acronymic syndrome of the topics, the vernacular and the hype. The genetics SFB added much to the fame of Cologne's science. When it ran out after its allotted and twice prolonged time, it was replaced in the same direction under a changed name.

The spring courses were replaced soon by the meticulously planned and executed "Spring Meetings" that gathered on the stage eminence and internal discussion before a mostly silent plenum. Unfortunately, I often did not have the time to attend, and I regret it now in my old age. It would have added to my knowledge of material, man and k.o. mice.

From provisional to solid fundament on fortified ground

Genetics, be it white, green or red, grew branches and with them the Genetics Institute grew. It had to be enlarged, later annexed, and, finally, completed on what will be the *"Biologicum-in- spe"* (under construction in the trenches of the Prussian Fort 5). When the Institute of Biochemistry was moved into the new building connected at an angle with the Genetics Institute, I thought it a splendid, productive idea that there were no doors in between and common contact rooms with coffee-machines and chairs to exchange extramural interests and ideas, rather than only exchanging methods at the bench. It did not take long until walls with doors and extra keys were installed for separate seminar or computer rooms instead of the well-used kitchen-lines. Formerly, genetics was an "open house", just as my Bottmühle office that was never locked. Now, in the new surroundings, even the hall has become so narrow that it is hard to find a place (on special permission) for my bike. And, since dealers nevertheless roamed around and stole my PC, symbol of conversion — and my logo-type vehicle — I try not to forget the *New Deal* of Open Science.

17

History and Fate of a Similar Concept: the Biology Division of the Southwest Center for Advanced Studies in Dallas

Hans Bremer

This is a story about the rise and fall of the Biology Division of the Southwest Center for Advanced Studies in Dallas, which can be regarded as an offspring of the Institute of Genetics in Cologne. In the early sixties, the electronics firm Texas Instruments found it difficult to hire physicists willing to move to Dallas, and for that reason, they decided to "produce" their own Ph.D.s in a research facility to be newly developed. To head this new facility they appointed the geologist Lloyd Berkner, who had been president of the International Geophysical Year in 1957, when the Antarctic was divided up among several countries. Berkner joined with another geologist, Larry Marshall, who was to hire the first faculty for this institute. Larry Marshall had the idea that pesticides would gradually kill the blue-green algae in the oceans and he predicted a catastrophic disruption of the oxygen cycle in the atmosphere, so that at some time in the future all life on earth would suffocate. Because of these concerns, biology became the largest division at the new institute.

Larry Marshall had approached Max Delbrück for advice and was referred to Carsten Bresch in Cologne and Günther Stent in Berkeley, who, in turn, suggested Roy Clowes in London, as potential candidates to head the new biology division. Between 1964 and 1968 Carsten Bresch organised the department, often commuting between Dallas and Germany, and when he left Dallas in 1968, Roy Clowes took over as the head of the department.

In our Biology Division, two areas of research were predominant: the study of bacteriophage (mainly phage of Delbrück's T-series), and the damaging effects of ultraviolet (UV) radiation on nucleic acids. In the photograph taken in 1968 (see Fig. 24) we see Carsten Bresch, who studied phage T1, Rudi Hausmann studying phage T3 and T7, and Yvonne Lanni and Jim McCorqudale studying phage T5. The UV effects on DNA were studied by Stan Rupert, who had discovered UV-induced thymine dimers in DNA and their repair by "photo-reactivation". He collaborated with Walter Harm (from Cologne), and Mike Patrick. John Jagger studied effects of "near-UV" on bacterial growth, and Harold Werbin, a biochemist, purified and studied DNA repair enzymes. The electron microscopist Dimitrij Lang (from Kleinschmidt's laboratory in Frankfurt) had shown that the DNA of phage T4, with over 100 genes, consists of a single, very large

Fig. 24. Division of Biology, Southwest Center for Advanced Studies, 1968, Faculty. From left to right: Jagger, Bresch, Patrick, Hausmann, Harm, Bauerle, Werbin, Rupert, Lanni, Gutz, Bujard, Hirokawa, Clowes, McCorquodale, Krone, Harris, Lang.

linear molecule. This was at a time when it was still speculated that every gene in a cell might be represented by a separate DNA molecule. Working with Dimitrij was Hermann Bujard, who studied *Papilloma* virus DNA and single-stranded *fd* phage DNA. The picture also shows Herbert Gutz, a yeast geneticist, Hideo Hirokawa, studying *B. subtilis* transformation, Winfried Krone, a human geneticist, and Dan Harris, an enzymologist. We were 21 professors, each working, on average, with one post-doc and two technicians. In addition, we had 10 secretaries and many other staff members, all financed by the Texas Instruments Corporation. All in all, we were an optimistic group with much social interaction and scientific discussion, and constantly planning the hiring of new faculty.

Naturally, this extravaganza could not last forever, and already a year after this picture was taken, Texas Instruments announced that they could no longer fund us. This started a new period for the department. But before I come to the end of it, I would like to lead you all the way back to the beginning of molecular biology.

In 1952, Carsten Bresch and Thomas Trautner, perhaps the first phage scientists in Germany, were working at the Max Planck Institut für Physikalische Chemie of Karl Friedrich Bonhoeffer in Göttingen. At the time I was a graduate student in the Zoology Department in Göttingen, studying the development of blood vessel patterns. My supervisors, Karl Henke and Joachim Autrum, thought that the development of arteries and veins in our body might be preceded by changes in pH and redox potentials, and they had sent me to the Bonhoeffer Institute to build suitable microelectrodes to measure redox potentials in living tissues.

The Bonhoeffers and Delbrücks had been neighbours and friends in pre-war Berlin, and after the war, Delbrück frequently came from Pasadena to visit Bonhoeffer in Göttingen. I remember when he gave a seminar there in 1953 about the "invention of the *Herren* Watson and Crick". Delbrück concluded that, on replication, there must be one break in the DNA every ten base pairs to allow separation of the two daughter strands, an idea that was later abandoned. In 1954 Delbrück visited again and gave the first molecular biology course in

Germany, in the Göttingen Zoology Department, where he explained his aim was to "rewrite page one of future biology text books".

By 1958, I had finished my dissertation (no redox potentials were involved), and my supervisor, Karl Henke, had died. Carsten Bresch had received his *Habilitation* from the University; his *Habilitationsschrift* was given to me to read because nobody in the zoology department could quite understand it. Bresch and Trautner had moved to Cologne, where Delbrück's new Institute of Genetics was to be built, and I applied for a post-doctoral position with Bresch. A few months later I was working in a little house in the Gyrhof Strasse in Cologne, next to the Botanical Institute, that had been converted into a primitive microbiology laboratory. The bathroom of that house had become a radioactivity lab, where Peter Starlinger taught me to label RNA with radioactive P^{32}. We wanted to determine the rate of messenger RNA synthesis in *E. coli*.

Our experiments did not succeed, because we did not realise that the turnover of the mRNA causes a reduction in the radioactivity of the nucleotide precursors in the cell that makes it impossible to determine the rate of mRNA synthesis in this manner. For this reason, I had no publication from my work in Cologne, and I had the honour of being fired by Max Delbrück. Fortunately, I was saved by Günther Stent, a friend of Delbrück and Bresch, who came to visit Cologne in 1961 after a sabbatical with Sydney Brenner in England.

In England, Stent had studied what came to be known as the "stringent" and "relaxed response" of *E. coli* RNA synthesis to amino acid starvation. To continue this work, Stent had to map the *rel* gene, whose mutation was responsible for the "relaxed" phenotype of his mutant. For that purpose, he brought Roy Clowes, a leading expert in bacterial genetics, to Berkeley. At the same time he wrote to Carsten, that he would like a certain person from Carsten's laboratory to join him in Berkeley, but he could not recall that person's name. They agreed that Carsten should send pictures of all his co-workers to Berkeley. To everybody's surprise, the answer came that it was me he wanted. So from 1962 to 65, I was studying *in vitro* transcription of phage T4 DNA in Berkeley, together with Stent's graduate student

Mike Konrad. We determined the properties and stability of the elongation complex, the direction and velocity of RNA chain growth, the initiating nucleotides, RNA polymerase binding to the DNA, and other things which are common knowledge today but were new at the time.

In 1964 Delbrück came to Berkeley, speaking again about bacterial DNA replication, but this time his talk was based on an auto-radiographic picture by John Cairns, which showed the *E. coli* chromosome to consist of a single, cyclic DNA molecule. Delbrück then proposed that there has to be a swivel at the origin of replication at which the DNA rotates at a speed of 10.000 rpm to unwind the double helix during replication (this idea had also to be given up later). When told by Günther Stent about my experimental progress, Delbrück rehired me for the Cologne Genetics Institute, but by then I had already decided to take Larry Marshall's offer to join the new institute in Dallas.

In 1968, Carsten Bresch and most of the German faculty he had brought to Dallas went back to Germany. This became known as the "exodus". A year later, in 1969, Texas Instruments ended their support and donated our research facility to the University of Texas: we became the new campus "UTD", the University of Texas at Dallas.

The new administration told us that "molecular biology" would be bad for public relations because we were known to create "Frankenstein monsters". We were told we should stop all that phage work. We also learned that (in Texas) research was considered a luxury for lazy professors to avoid teaching. From now on we should concentrate on teaching biology. For many Americans, this means to dissect a cat in a high school class. Despite their contempt for research, they wanted to become the "MIT of the Southwest" (the current idea to create an "MIT of Europe" seems to reflect similar desires).

Thus began our never-ending uphill fight with the UTD administration. However, our department was able to attract excellent graduate students and it became the largest and most active "Program" at UTD ("Departments" were abolished "to keep budgetary power high in the administration"). When Roy Clowes died in 1987, the decline was unstoppable and today the department has almost totally collapsed. I am glad to hear that its mother institution here in Cologne has experienced a better fate.

To end, I believe I should address some problems of the future affecting all university life in the United States. One problem is the enormous grade inflation, which has made grades or grade point averages meaningless for most post-university careers. Many colleges state that they are looking for students with "leadership skills" and talents other than academic excellence. To preserve at least some academic excellence, it becomes necessary to have "elite" universities, apart from the numerous "average" universities. Another problem, which hinders the economic development of poorer countries, with many undesirable consequences, but greatly benefits the United States, is the continuing brain drain from those poorer countries. Finally, there are now wide-spread political accusations that universities have become too far left. In the last presidential election, 10–20 times more professors at the top universities supported Kerry rather than Bush, and it is asked: "How did academia, supposedly dedicated to the free exchange of ideas, become so intellectually monochrome?" This emerging rift between "rightist" spokesmen in the press, often associated with the business lobby, and "leftist" intellectuals needs to be addressed to create a healthier climate for university education.

18

TMV in Tübingen and its Escapade with Genetics

Karl-Wolfgang Mundry

I was really surprised when I received the invitation to participate in this meeting and I want to thank the organisers for the opportunity to be here today. It is an honour for me to tell you about my years in Tübingen, where I worked with Tobacco Mosaic Virus (TMV) about half a century ago. I am especially honoured since so many of you are distinguished geneticists and molecular biologists. What could TMV contribute to genetics? TMV is a plant virus containing single-stranded RNA instead of DNA. The RNA does not recombine and has a terrible plating efficiency of about one in a million. TMV requires large tobacco plants for handling in virus-sterile greenhouses instead of Petri dishes, while *Escherichia coli* can be handled with a lab bench and an incubator. Despite these handicaps, TMV was — and still is — a pioneer object in molecular biology, but did it play a role in genetics?

At the very start of what later became molecular biology, an idea was born: could viruses some day serve as model systems for the investigation of replication processes and hereditary mechanisms? In our country, this began in Berlin in 1938, when three directors of the Kaiser Wilhelm Institutes for Biochemistry and for Biology, Adolf Butenandt, Fritz von Wettstein and Alfred Kühn decided to install a working group for the study of viruses. To start virus work they delegated three scientists from their staff: Georg Melchers (father of Fritz), a botanist with a foible for genetics; Gerhard Schramm, a biochemist; and Hans Friedrich-Freksa, who was interested, like the others, in problems of general biology, with a foible for theories.

These three people began to work in 1938. They took advantage of two features of TMV: availability in rather large amounts, thus permitting chemical and physical studies; and what appeared to be genetic diversity. They stated their aims very clearly in an introduction to a paper published as early as 1940, where they wrote: "Since plant viruses are known to be nucleoproteids of defined composition it is an important task for virus research to deduce differences in the biological behaviour of virus species to differences in their chemical set-up". When their institutes were evacuated from Berlin during the bombing in the second world war, they settled in Tübingen and took their TMV research with them. Thus Tübingen in its Melanchthonstrasse became one of the three centres of TMV research (as well as for some animal viruses) in the world. But Tübingen also became the place where TMV was studied, to a great extent, under genetic aspects and so entered an "escapade" (*einen "Seitensprung"*) into genetics.

My relation to TMV began along these lines. As a student of botany I asked my professor in Göttingen, Richard Harder, a distinguished scholar, in 1948: "What is a virus?" He admitted not knowing the answer and continued: "I wonder if anyone anywhere already has the answer to your question, but if there is someone in our country who can tell you something about it this may be Herr Melchers in Tübingen." So I took my bicycle, travelled 600 km to Tübingen and got a two-day interview with Melchers. A year later, after his new institute was finished, I got the chance to work with him. "I want you to study the problem of whether TMV can be made to mutate," he said and added: "We shall later see, whether it will become a thesis!" This was in the fall of 1950.

Before I continue, I should acquaint you with the research object. Tobacco plants, when infected with TMV, produce symptoms of a disease. Symptoms on the experimentally inoculated leaves are called primary symptoms, while secondary symptoms are those which develop on the top of the plants while growth continues. Primary as well as secondary symptoms depend on the virus isolate used and are obvious genetic characters of the particular virus. For example, the secondary symptoms of a green strain are different from a yellow strain. Localised necrotic lesions are often primary symptoms and develop when leaf tissue reacts hypersensitively. The number of lesions is a function of the virus concentration and allows titration measurement of the infectivity similar to the plaque tests with phages on a layer of *E. coli.*

After weeks of introduction to the techniques of TMV research I started trying to mutagenise TMV by irradiation with X-rays. The tests made use of a tobacco variety which contained a particular gene N′ that permitted detection of local lesion mutants against a large surplus of non-necrotic wild-type virus (I shall come back to this test system later). Without doubt such mutants do arise spontaneously. In preparations of the non-necrotic strain, *vulgare,* 0.2% of the infectious material is of the local lesion type; you cannot get rid of this "contamination". To make a long story short: I failed completely. Although highly concentrated samples of TMV had been inactivated by irradiation of six orders of magnitude, there was not the slightest effect in the mutation rate of the virus. Efforts to mutagenise TMV with UV instead of X-rays remained equally unsuccessful.

In vitro mutagenesis with X-ray or UV-irradiation did not work (understandable nowadays because of the missing repair mechanism for RNA). But could the temperature at which the host plants grow influence the mutation rate *in-vivo*? There was a report from 1935 from an experimental tobacco station in the United States which stated that tiny yellow spots, which appear occasionally within the green-type mosaic of TMV, seem to increase in number during hot summer seasons. From these spots "yellow strains" of TMV may be isolated. Is the mutation rate of TMV temperature dependent during its replication in plants? The answer is: Yes! In a number of experiments it became clear, that the higher the growth temperature, the more plants exhibit such tiny yellow spots. This happens when the infection invades younger, not experimentally inoculated leaves of the plant. In addition: the number of yellow spots per plant also increases with the growth temperature, while the production of TMV exhibits a clear optimum around 27°C. When the number of yellow spots per plant is related to the amount of virus produced per plant, the mutation rate of TMV increases only if the temperature exceeds the optimal condition for virus production.

I wanted find out whether selection processes could simulate what seemed to be an effect of temperature on the *in-vivo* mutation rate of TMV. For this purpose I did a temperature pulse experiment: Sets of 21 plants were inoculated at time zero and exposed to 35°C for 24 hours at the first, second, third, fourth or fifth day post inoculation. The rest of the time they were kept in the greenhouse at about 22°C. The number of yellow spots per set of plants decreases with the age of the infection. This means, that the TMV mutants within the yellow spots on the plants cannot compete with the wild type during the invasion of leaf tissue when they leave the veins. It also means that they only have a reasonable chance to express themselves if they hit uninvaded tissue (that is at the beginning of secondary infection).

However, Melchers was satisfied. I got my Ph.D. and moved to Braunschweig, where I accepted an offer as head of the small virus department at the Institute for Agricultural Technology and Sugar Industry at the Technische Hochschule [today: Technical University Braunschweig]. There I had to work on the identification and characterisation of sugar beet viruses, which caused a severe reduction of the yield of sugar beets in the fields at that time (in some regions below 20% of the average harvest). Nevertheless, I continued to do some work on the interference of different TMV strains. This permitted, after mutagenesis had become successful, a careful exclusion of possible selection phenomena.

Three years later Melchers offered me a position at the Max Planck Institute for Biology in Tübingen, obviously believing I had suffered enough of what had been in his mind "scientific diaspora". I came back to Tübingen and into an extremely exciting intellectual atmosphere. Alfred Gierer and Gerhard Schramm in our neighbouring institute, the Max Planck Institute for Virus Research, had just isolated an infectious molecule for the first time in science, the RNA of TMV. Gierer had shown later, by inactivation kinetics with ribonuclease, that the infectious molecule is single-stranded, again a "first" in science.

Meanwhile, Heinz Schuster and Schramm had been playing with several procedures to reliably inactivate viruses. The first large-scale routine vaccinations of children in the United States with formaldehyde-inactivated poliovirus had resulted in a number of grave mishaps. Among other possibilities, Schuster and Schramm studied the effect of nitrous acid (a slightly acidic solution of sodium nitrite) on the infectivity of TMV-RNA; as Schramm and Müller did previously in 1942 with TMV nucleoprotein. Nitrous acid converts the amino bases of nucleic acids into oxy bases, namely C, A, and G into Uracil, Hypoxanthine and Xanthine, respectively. Heinz Schuster and Schramm found, surprisingly, that TMV-RNA can be inactivated by orders of magnitude while the integrity of the RNA molecule remains intact. They also found that the inactivation follows first-order kinetics, meaning that deaminating a single base is sufficient to render the RNA molecule non-infectious. According to some quantitative calculations, about one-third of all amino bases in the RNA of TMV appear to be resistant to this inactivating effect, indicating that these, if they were "hit", could — perhaps! — bring about mutations. One fascinating aspect was that in the case of conversions of cytidin to uridin residues one natural constituent of RNA would replace a deaminated base. After several unsuccessful trials to find mutants we switched to the Java tobacco system. I had used that system before in my irradiation studies, where I looked for local-necrotic mutants of the non-necrotic wild-type TMV *vulgare*. Though it was quite clear that such mutants do arise spontaneously, all of us remained very sceptical. The greater was our excitement, when three days later the leaves of the test plants were covered with hundreds of lesions, and collapsed completely within the two following days.

The effect was so dramatic, that we first thought we had mixed up bottles and had used a necrotic variant of TMV, the strain dahlemense, instead of the systemic wild-type *vulgare*. But everything was reproducible, and after an intermission of several months, because I had broken a leg while skiing in the Austrian Alps, kinetic analyses under milder conditions showed without any doubt that the percentage of mutants among the survivors increases appreciably. We also showed that the mutagenesis follows first order kinetics. In addition, since there was no difference in the behaviour of native TMV and isolated RNA, the protein part of the virus did not play a role. Since possible selection problems could be ruled out on the basis of my experiments in Braunschweig, the "atom of mutability", as a journalist called it, had been identified.

This finding opened up perspectives to participate in the efforts to break the genetic code. When we (Gierer and myself) presented our single-hit nitrous acid mutagenesis at the fifth International Congress of Biochemistry in Vienna in 1958, the very first question came from Vernon Ingram, the discoverer of the single amino acid difference between sickle cell and wild-type haemoglobin. He asked: "What is the influence of your mutagenesis on the capsid protein of TMV?" I answered: "We do not know it yet, but this is under investigation by Heinz Günther Wittmann at our institute." Wittmann's first comparison between wild-type capsid protein and that of a mild mutant (producing almost no visible symptoms on tobacco, a mutant from our single-hit collection) did not reveal any difference in amino acid composition of the different RNA base sequence. However, within an admirably short period of time he and his wife Brigitte Wittmann-Liebold demonstrated convincingly that *in vitro* mutagenisation of TMV with nitrous acid causes amino acid substitutions in the capsid protein of the virus. They also accumulated overwhelming evidence that most, if not all, substitutions they found do not only agree with the code words as identified with the *E. coli in vitro* systems of Nirenberg and others, but also that their data lent strong support to a degenerate triplet code, with two nucleotides more important than the third. Their results contributed to the view that the code is probably universal.

The strategy of this work is shown in a scheme: starting with polycytidylic acid (known by then to code for polyprolin) and polyuridylic acid (coding for polyphenylalanin), deamination of an RNA messenger with nitrous acid should, at least occasionally, cause the substitution in its polypeptide product of a prolin by a phenylalanine. However, with a triplet code this could not result from a single deamination event. Instead — and this is one main achievement of the Wittmanns' work — phenylalanine appears in the position of a prolin residue in a polypeptide chain only via one of two intermediates: serin or leucin. This (and dozens of additional findings) was in full agreement with what had accumulated between 1960 and 1965 from the work of a number of groups to be the structure of the genetic code. Specifically, the code is a triplet code, is non-overlapping (because a single deamination had never caused the substitution of two adjacent amino acids!), is degenerate, and it is probably universal. The universality was shown because the code word assignments obtained with a prokaryotic *in vitro* system from *E. coli* matched completely those obtained with the eukaryotic *in vivo* system of TMV-infected tobacco plants. And that was TMV's escapade into genetics.

V. Research and Scientific Collaboration

19

Molecular Virology and Medical Genetics at the Institute of Genetics in Cologne, 1972–2002

*Walter Doerfler**

Prelude

As the Institute of Genetics in Cologne approaches its 50th anniversary, we have been asked to reminisce about our experiences in the Institute during its early period. My report will cover the period between 1972 and 2002 and will necessarily come across as a very personal account.

After completing Munich Medical School in 1958 and a doctoral thesis in functional anatomy with Titus von Lanz in 1959, I became medically qualified (*Approbation*) after completing about three years of clinical work at Munich Medical School and a one year rotating internship as a fellow of the Ventnor Foundation at Mercer Hospital in Trenton, New Jersey, in the US.

Adolf Butenandt's lectures in Munich had fascinated and motivated me to seek more extensive post-doctoral training in biochemistry. I was fortunate to be accepted as a post-doctoral scientist in Wolfram Zillig's laboratory at the Max Planck Institute for Biochemistry in Munich between 1961 and 1963. Wolfram Zillig became a fantastic

*In memorium Walter Kreuter, 1877–1952.

and inspiring teacher and friend. The early 1960s, of course, were most exciting years in what later became known as molecular biology. With my earlier positive experiences in the United States, it was only a natural step to apply for a second post-doctoral period there. With Wolfram Zillig's encouragement, I applied to the Department of Biochemistry at Stanford University — and was accepted by David S. Hogness.

First visit to Cologne — course in bacterial and phage genetics

Now it was high time to study genetics, because Dave Hogness' expertise was clearly in a combination of robust DNA physical chemistry and bacteriophage λ genetics. In 1961, Max Delbrück had founded the Institute of Genetics in Cologne as a pioneer institution in Germany, not only in genetics but also as an example of how a university department could be organised. In March of 1963, I made my debut visit to the metropolis on the Rhine and to the new Institute. The course, taught by Carsten Bresch, Peter Starlinger and Thomas Trautner was an excellent experience.

Reminiscences about Max Delbrück

Of course, I knew that he was one of the founders of Quantitative Molecular Biology. He had given a lecture at the Max Planck Institute of Biochemistry in Munich that we all attended. But now in Cologne, there was a chance to interact with Delbrück personally, for instance during a Sunday morning *Spaziergang* along the Rhine followed by a *Borscht* soup luncheon in the Delbrück home in Cologne. I had heard stories about Max Delbrück and decided to pretend to meet him *en par,* an exaggeration certainly on my part, although I managed well with that approach.

In the spring of 1966, during my time at Stanford, I visited Max and Manny Delbrück in their home in Pasadena. I told Max Delbrück about my plans to join the junior faculty at Rockefeller University later that year and to start working on DNA Tumour Viruses. "You will not be alone," was his dry prediction that, of course, turned out to be very true.

During one of Delbrück's visits to Cologne in the 1970s, my wife Helli and I invited him home for dinner. When I took him back to the hotel, he confided that in his years in Berlin in the 1930s, as a physicist being interested in the phenomenon of life, he and his colleagues had speculated about a fifth elementary force in physics that might explain the events facilitating life on earth. This new principle they did not find and were disappointed to learn that good old physics and chemistry seemed again to account for most of the mechanisms active in biological processes. While so many astounding new ideas are discussed today in physics (4% "normal" matter, 26% dark matter, 70% dark energy), perhaps we will not be too surprised if Delbrück and his colleagues would not have been too far off the mark after all.

During another visit to our home, Max Delbrück got into a conversation with our then eight year old son, Markus. "Get me a deck of cards," he suggested and then a most amazing card game took place; Markus being asked to guess the colours of the cards Max Delbrück drew from the carefully shuffled deck. In ten to fifteen exchanges Markus always guessed right. Delbrück exhibited the face of a sphinx and did not answer my questions about the trick he had applied. I will always remember Max Delbrück as an uncompromising, quite natural, and, in his way, friendly person, with an eminently interesting, inquisitive and demanding mind. It was one of my great experiences to have known Max Delbrück, not only as a great scientist but as a real person.

Department of Biochemistry, Stanford University Medical School — post-doc with David S. Hogness

On 1 September 1963, I arrived in the department, and Dave Hogness and his colleagues were to become my most important educational experience. Most molecular biologists would probably have agreed that this department, shaped by Arthur Kornberg, Paul Berg, Israel Robert Lehman, Robert Baldwin, Dale Kaiser, Lubert Stryer, George Starck and, of course, David S. Hogness, was the best of its time.

With Dave as a fantastically critical mentor, we managed to physically separate the complementary chains of bacteriophage λ DNA, to construct heteroduplex molecules between the strands of wild-type λ DNA and those of a home-made double mutant, λ N_7N_{53}. Before transfecting the heteroduplexes into *E. coli,* its repair mechanisms had to be sequestered by UV irradiation of the recipient cells to find differential activities of the two complementary heteroduplex DNA molecules. In this way, we could identify the transcribed DNA strand for gene N in bacteriophage λ.

Seminars galore and many encounters with fellow scientists from around the world provided a first rate education in Molecular Biology. These were happy days in Palo Alto, California in our personal lives as well. Helli enjoyed life in the San Francisco Bay area and work in the Department of Pediatrics with Dr. Sussman until our son Markus was born on 21 May 1965, the happiest day of them all.

Rockefeller University, molecular virology, Igor Tamm, Purnell Choppin

One of the visitors, in the Stanford Department of Pharmacology, was Igor Tamm from Rockefeller University. We had an extended discussion during that visit on the possibility of my joining the Rockefeller Virology laboratory as an Assistant Professor after Stanford. I gave my first seminar at Rockefeller on 7 December 1964 and started the RU appointment on 1 April 1966. Igor Tamm at many occasions was a generous mentor and department chair and helped us to facilitate the transition from Stanford to the quite different, but no less exciting, life in New York City. Rockefeller University, of course, was a perfect match for Stanford with regards to the intellectual and scientific atmosphere. Never before or after did we experience that spirit which pervades New York City at its best, particularly as exemplified by its cultural life on and off the Rockefeller campus.

While still at Stanford, I had made up my mind to work on adenoviruses and to investigate the fate of the adenoviral DNA upon infection of cells and the induction of tumours by adenovirus type 12 in newborn hamsters. Igor Tamm not only accepted but supported my intentions. I also wish to acknowledge the help from Walter

Schlesinger and his group at Rutgers University in Piscataway, New Jersey. In 1968, I published the first data showing that adenovirus type 12 DNA became covalently linked to cellular DNA in abortively infected hamster cells. I had also started collaboration with Albrecht Kleinschmidt's group at New York University on the structure and the denaturation profile of adenovirus DNA. The work on adenovirus molecular biology had attracted a small group of researchers: Ulla Lundholm (from Stockholm), Byron Burlingham (from Iowa), Monika Hirsch-Kauffmann (from Cologne), and Harold Burger (from New York). Lennart Philipson from the *Wallenberg Laboratoriet at Uppsala Universitet* came on visits to the Virus Group as an alumnus of that group. We soon became involved in a joint project with Ulf Pettersson, Byron Burlingham, Lennart and myself on an endonuclease activity associated with the penton subunit of the adenovirus particle. I really enjoyed working at Rockefeller University and we liked the big city a lot.

Alternative offers

Several offers appeared: from the University of Miami; from the Downstate Medical School, State University of New York; from Yale University (which got me the promotion to Associate Professor at RU in 1969); from the Max Planck Institute (MPI) in Tübingen, and later in Berlin (both as a junior group leader); from the MPI in Berlin also in a position similar to a C3 university job; and, finally, one for a Chair in Microbiology, formerly held by Hartmut Hoffmann-Berling, at Heidelberg University in 1969.

In August 1970, Peter Starlinger from Cologne visited the laboratory at Rockefeller, and he told me about their exciting work on insertion elements in *E. coli*. A few months later, I received an invitation from Peter Starlinger to talk at probably one of the first Cologne Spring Meetings, in 1971. After my presentation, Peter Starlinger, Klaus Rajewsky, Benno Müller-Hill, Walter Vielmetter and Peter Overath asked me how firmly I was committed to accepting the offer in Heidelberg. It is difficult today to convey the problematic situation with the administration and the faculty of Heidelberg University in 1970/1971. So it was a pleasure and relief to get an offer from Cologne with a Sonderforschungsbereich of the Deutsche Forschungsgemeinschaft to be newly established, and an Institute structure that was pervaded by the Delbrück spirit and was at least vaguely similar to a department in at a university in the U.S.

Aside from the attraction of the group at the Institute of Genetics, the Delbrück legacy, and the metropolis on the Rhine, discussions in the spring of 1971 with Willy Stoffel, professor of Biochemistry in Cologne, and himself an alumnus of Rockefeller University, were convincing and helpful. So, I felt ready to jump, but did have a safety net in place: Igor Tamm and Fred Seitz, the president of Rockefeller University in 1971, had offered me the honorary position of Adjunct Professor at Rockefeller that I held until 1978. It was also said that I could always come back to the campus on the East River. Moreover, Lennart Philipson had asked me to come to Uppsala University as Visiting Professor for the period of 1971/1972, with support by the Swedish Cancer Society, and consider possibilities in Uppsala as well.

Wallenberg Laboratoriet vid Uppsala Universitet

In September of 1971, with much hesitation, I left Rockefeller and started to work in Uppsala. The atmosphere in the laboratory was stimulating, and I learned a lot from Lennart Philipson's set up about running a laboratory with many young students. One of the lasting experiences was the many scientific contacts and personal friendships I could develop with my colleagues in Uppsala. During later years I returned for frequent visits, sometimes as the opponent for doctoral dissertations; upon the invitation of Ulf Pettersson for Aslin Tulan and Catharina Hellström, and for Janos Minarovits, doctoral student of George Klein and Ingemar Ernberg at Karolinska Institutet in Stockholm. In 2002 and 2006, we have started new collaborative projects with Ulf Pettersson's group at *Uppsala Universitets Rudbecklaboratoriet.*

And now on to thirty years at the Institute of Genetics in Cologne

All equipment and most of the financial support for the new virus group at the Institute of Genetics had to come from the DFG. The university made the formal offer for the C4 position, space and a number of university positions. Funds for an adequate start-up (*Grundausstattung*) at the level required could, however, not be procured. As the DFG-supported offer was quite generous — it was linked to a five year DFG professorship which I had originally accepted — I did not worry any longer about the implementation of the university offer. Without the support by the DFG, I would never have accepted the position at the Institute of Genetics.

The concept of research-oriented teaching is a given at all research universities in the US, and nowadays has also been implemented at many university institutes in Germany. The Institute of Genetics, while firmly grounded in the University of Cologne, was a research institute with the aim to involve students early on in their education in internationally competitive research. The interest of young students in this discipline was encouraging, and breath- and space-taking at the same time. The first course in advanced genetics that I participated in teaching was taken by 24 students. Within only a few years that number swelled to over 120. It would be fair to say that our generation of researchers in Molecular Biology imported the discipline to Germany in the 1960s and 1970s. A whole generation of students and post-doctoral researchers from Germany and many other countries was educated during these and earlier years at the Institute of Genetics in Cologne.

The beginning in Cologne was graced by a streak of good luck. The group (see Fig. 25), not only of students but also of post-doctoral researchers and independent investigators, that I managed to convince to join the laboratory in Cologne came from many different countries and helped to continue the international atmosphere I had so much become accustomed to during the preceding decade. Harold Burger, my second doctoral student from Rockefeller, was given permission to finish his work in Cologne; Ellen Fanning had come from Madison, Wisconsin; Rona Greenberg, a student from Berkeley and Lily Vardimon

Fig. 25. The Molecular Virology Group, Institute of Genetics Cologne, in 1977. *Top row*, from left to right: M. Fischer, S. Zeiger, G. Wollny, Lennartz, M. Westphal, B. Kirspel, M. Brötz, P. Kathmann, H. Esche, J. Schick. *Middle row*: W. Doerfler, B. Bode, D.T. Brown, R. Dunker, D. Sutter, B. Riedel, L. Vardimon, B. Weingärtner, R. Neumann, M. Stupp, U. Winterhoff. *Front row*: D. Renz, S.T. Tjia, J. Smith, K.-H. Scheidtmann, R. Schilling, K. Baczko, H. Mansi-Wothke, E. Fanning, E.-L. Winnacker.

from Jerusalem joined the group; Juan Ortin, who had earned his Ph.D. in Margarita Salas' laboratory in Madrid, came as a post-doctoral researcher to our group; Diane Sutter from Pennsylvania as a doctoral student; Sian T. Tjia was a student who had left Surabaya, Indonesia and was to become my trusted and most helpful colleague, and friend in the group for the next 30 years; and Monika Westphal could be persuaded to lend her expertise gained during her years at Rockefeller to our group.

The *Sonderforschungsbereich* support also enabled me to attract two outstanding independent investigators to the group: Dennis Brown, a top echelon virologist and electron microscopist (trained by Thomas Anderson in Philadelphia), came from Baltimore; and Ernst-Ludwig Winnacker, a superb biochemist who had trained with Eschenmoser in Zürich, Barker at Berkeley, and Peter Reichardt at the

Nobel Institute, came from Stockholm. This enthusiastic group stayed together until 1978, when the first students had graduated and the independent investigators received offers that Cologne failed to match. The good luck in attracting outstanding international investigators continued in later years, but space limitations do not allow me to mention all our later group members.

Between 1967 and 2006, I had the privilege to personally guide the work of 80 graduate students, including two at Rockefeller University. Among them several (about 20%) came from other countries: USA (4), Israel (2), Indonesia (1), Turkey (1), France (1), China (1), Poland (1), Iran (1) and Brazil (1). I also supervised the research of more than 100 diploma students, and counselled about 40 post-doctoral researchers. Twelve young researchers reached the *Habilitation,* a university exercise that likely will, and may as well, vanish in the future. And where did they go? Research in academia, 31%; corporate research, 25%; management (academia and corporate), 24%; and medicine, 2.5%. More than 30% of our former associates work outside Germany.

In Arthur Kornberg's stimulating book, *For the Love of Enzymes* (1989), he states, "If the case can be made that my activities in administration, writing, and teaching have made a unique contribution, then certainly a further case can be made that my discoveries in science have not. Very likely, they would have been made by others soon after." If Arthur Kornberg, a role model scientist and teacher for many of us fortunate enough to have known him, assigns that importance to teaching, writing and science administration, we are well advised to place our own emphasis accordingly. Of course, the level of scientific education achieved by the Department of Biochemistry at Stanford could not have been achieved without a world-class research programme in this department. The concept of research-oriented teaching cannot be recommended highly enough for any university education.

Space problems

The Institute of Genetics had been constructed in 1961 with the then substantial, but in actual terms quite modest, sum of DM 3.5 million.

The building, while adequate for science in the early 1960s, soon proved to be too small for expanding research in the five departments that had allotted some of their space to one or several independent groups. Newly available technology and safety regulations required additional space and, most importantly for a university, the number of students, who became aware of the "revolution" in biology, increased precipitously every year. Fortunately, in 1981 a small addition to the building on Weyertal was financed by the Volkswagen Stiftung and alleviated the most pressing problems in the Institute. Moreover, at that time plans appeared on the far away horizon that the University would invest in a *Biozentrum* — as it was to be called — at a different location on university grounds. In 1982, the faculty of the Institute of Genetics submitted its first application for a larger Institute. In 1991, the first instalment was completed on Zülpicherstrasse, about ten minutes walking distance from the old Institute. *Das Neue Institut* was finally completed and officially opened in 2006 with a new faculty. We are all delighted that after almost a quarter of a century our plans of 1982 have materialised.

Major research topics in molecular virology/medical genetics at the Institute of Genetics in Cologne

1. *Adenoviruses as Tools for Studies on the Molecular Biology of Mammalian Cells*
   - (i) Structure of the core of the adenovirion (in collaboration with Dennis T. Brown)
   - (ii) Symmetric recombinant of Ad12 DNA spontaneously arising in virions from productively infected cells
   - (iii) Is there also integration of adenovirus DNA in productively infected cells?
   - (iv) Endonuclease in adenovirus-infected cells (continuation of the collaboration with Lennart Philipson's laboratory. This collaboration was resumed in 2006 with Ulf Pettersson's laboratory in Uppsala)
   - (v) Adenovirus oncogenesis and cell transformation

(vi) Revertants of Ad12-transformed cells or of Ad12-induced hamster tumour cells

(vii) Integrated state of adenovirus type 12 (Ad12) or type 2 (Ad2) DNA in transformed hamster cells or in Ad12-induced hamster tumour cells

(viii) Development of an *in vitro* system to study recombination between Ad12 DNA and cellular DNA

(ix) Transcriptional activities of the adenovirus genome in the integrated state

(x) Loss of all integrated Ad12 DNA sequences from Ad12-induced hamster tumour cells is compatible with the maintenance of the oncogenic phenotype

2. *Human Adenovirus Type 12: Crossing Species Barriers to Immortalise the Viral Genome*

(xi) When viruses cross species barriers, they often drastically change their biological and pathogenetic properties. In my laboratory, the abortive interaction of non-permissive Syrian hamster cells with human adenovirus type 12 (Ad12) has been studied. Ad12 induces undifferentiated tumours in newborn hamsters (*Mesocricetus auratus*) at high frequency within 4 to 6 weeks. Ad12 inefficiently enters hamster (BHK21) cells, and only minute amounts of viral DNA reach the nucleus. Viral DNA replication and late transcription are blocked. In Ad12-induced tumour cells, multiple copies of viral DNA are chromosomally integrated. The integrated viral DNA becomes *de novo* methylated. Cellular DNA methylatiion and transcription patterns in Ad12-transformed cells and in Ad12-induced tumour cells are altered. These changes may be related to the oncogenic potential of Ad12 in hamsters. Thus, human Ad12 seems to be innocuous in humans but assumes efficient oncogenic potential after crossing the human-hamster species barrier.

3. *On the Functional Significance of DNA Methylation in Mammalian Systems*

(xii) Absence of 5-methyldeoxycytidine in adenovirion DNA (collaboration with Ursula Günthert and Manfred Schweiger in Berlin)

(xiii) *De novo* methylation of integrated adenoviral (foreign) DNA

(xiv) Inverse correlations between promoter methylation and gene activity, premethylation of promoter-indicator gene constructs, transfection and transient activity tests, studies in an *in vitro* system

(xv) DNA methylation patterns in different parts of the human genome

(xvi) Consequences of foreign DNA integration into established mammalian genomes

(xvii) Alterations of methylation and transcription patterns in mammalian cellular DNA upon the integration of foreign (Ad12, bacteriophage λ or plasmid) DNA

I consider this latter topic (xvii) of fundamental importance because its pursuance could shed light on unforeseen and unforeseeable problems arising during the generation of transgenic (gene manipulated) organisms, the cloning of organisms, and in gene therapeutic strategies; possibly also in knock-in and knock-out experiments that are so frequently the basis of medically relevant conclusions.

4. *Molecular Medical Genetics*

Starting in 1990, my laboratory extended its expertise on the genetic signal 5-methyldeoxycytidine, garnered from work on adenovirus- transformed cells, to the field of molecular medical genetics. At the same time, I started collaborations with several clinical departments, in particular the Institute of Medical Genetics at the *Universitätsklinikum* in Essen (with Eberhard Passarge, Gabriele Gillessen-Kaesbach and, at the molecular level, Bernhard Horsthemke). For several years, I attended their Clinic of Medical Genetics in Essen as a guest physician to learn also about their clinical aspects of some of the most important genetic diseases.

We investigated methylation patterns in several promoters of genes relevant for the following diseases:

(xviii) Prader-Willi-Syndrome; Angelman Syndrome (collaboration with Bernhard Horsthemke, Essen); Hirschsprung Disease (collaboration with Alexander Holschneider, Cologne and Giovanni Romeo, Genoa); Fragile X Syndrome (FRAXA) (collaboration with Alexander von Gontard, Cologne, Walter Vogel and Peter Steinebach in Ulm); the Waardenburg Type 4 syndrome (with S. Lyonnet, Paris); and Elliptocytosis and Spherocytosis (with Yoshi Yawata, Kawasaki

Medical School in Kurashiki, Japan). In the spring of 2006, I started a collaboration with Ellen Fanning's laboratory at Vanderbilt University in Nashville, Tennessee in the United States on methylation patterns in the promoter of the FMR1 gene, a region that also harbours an origin of DNA replication, in healthy and FRAXA individuals.

5. *Molecular Biology of Baculoviruses*

(xix) The work on this insect virus system started as a safety project to demonstrate that the insect baculovirus *Autographa californica* nuclear polyhedrosis virus (AcNPV) as a virus was not able to productively infect and replicate in mammalian cells. Later on, we became interested in the transcription patterns of this viral DNA in insect cells. Hermann Lübbert, one of the doctoral students, discovered overlapping transcripts in this viral system.

6. *Fate of Foreign DNA in the Gastrointestinal Tract of Mice*

(xx) The *de novo* methylation of integrated foreign DNA has been considered an ancient cellular defence mechanism. Apart from viral transgenes, most foreign DNA enters an organism via its gastrointestinal (GI) tract with the daily food supply. What is the fate of foreign DNA in the GI tract upon oral application? Upon feeding, small amounts (1%) of different test DNAs (bacteriophage M13, GFP, RUBISCO, Ad2 DNA) persist transiently in the GI tract as fragments of a few 100 to 1700 nucleotides long. Test DNA was followed by PCR, fluorescent *in situ* hybridization, recloning and sequencing. Test DNA could be traced to intestinal cells, peripheral white blood cells, the spleen and the liver. Transcription of persisting foreign DNA was never observed by RT-PCR. Test DNA appeared also in occasional fetal cells when pregnant animals were fed with test DNA. However, there was no evidence for germ line transmission.

7. *Parallels between Genetic and Linguistic Encoding*

Particularly in the 1980s, I pursued this interesting possibility and even tried to interest one of my colleagues in linguistics at the University of Cologne, Professor Seiler, in the problem. Although we both could recognise exciting parallels, we decided that our fields were not "far enough advanced" to base an interesting idea on sound scientific footing. (W. Doerfler. In search of more complex genetic codes — Can linguistics be a guide? *Medical Hypotheses* **9**, 563–579, 1982.)

Independent Group Leaders in the Department of Molecular Virology

Ernst-Ludwig Winnacker worked from 1972 to 1977 on the mechanism of replication of adenovirus DNA in productively infected human cells. He was the first investigator in our group whom I could help to obtain the *Habilitation*. Ernst Winnacker soon became professor of biochemistry at LMU in Munich, the founder of the *Genzentrum* there and, finally, the president of the Deutsche Forschungsgemeinschaft. As of 1 January 2007, Ernst has been appointed secretary general of the European Research Council.

Dennis T. Brown was an accomplished arbo-virologist and electron microscopist and joined our group from 1972 to 1978. In his own programme he followed the fate of Sindbis virus infections both in the permissive insect and the non-permissive mammalian hosts. He established many highly productive collaborations at the University of Cologne, also with my group. In 1978, Dennis became professor at the University of Texas in Austin, Texas in the United States. Today he is professor and chair of the department of Biochemistry of the State University of North Carolina in Raleigh, North Carolina in the United States.

Helmut Esche came to us in 1978 from Thomas Trautner's laboratory in Berlin as a post-doctoral fellow. After having spent several subsequent years at the Cold Spring Harbor Laboratory on Long Island in New York, he returned to Cologne as an independent investigator to work actively on mechanisms of transcription in the adenovirus system. He later became, and still is, Professor of Molecular Biology at the Universitätsklinikum Essen.

Silvia Stabel was an independent researcher at the Max Delbrück Laboratory in Cologne-Vogelsang that was supported through the Genzentrum Cologne, and was supported by researchers both at the Institute of Genetics and at the Max Planck Institute for Plant Molecular Biology (*Züchtungsforschung*). Silvia Stabel had done her dissertation in our group, then joined Lennart Philipson's laboratory at the European Molecular Biology Laboratory in

Heidelberg. She later worked with Dr. Parker's group at the Imperial Cancer Research Fund in London. When she was offered a professorial position at the University of Cologne, however, she decided to pursue a career as independent artist specialising in paintings with motifs from molecular biology and beyond.

Dagmar Knebel-Mörsdorf did her doctoral and post-doctoral studies in my group and then started her own projects on baculoviruses for which she defined a completely independent approach. When Dagmar Mörsdorf left the Institute of Genetics, she responded to an offer from the medical faculty of the University of Cologne where she now works as professor on problems related to baculovirus and herpesvirus entry into permissive cells. Dagmar has also proved to be a much appreciated university teacher and recently received an award from our Institute for excellence in teaching.

Sian T. Tjia, who had pioneered the baculovirus system in our group as a doctoral student, later instructed a large number of national and international visitors in the biology of that virus. As the AcNPV system had become one of the very useful eukaryotic expression systems, many researchers from many different universities, research institutes and from industry from many different countries visited our laboratory to study the technical aspects of this system. Moreover, together with Petra Böhm and Susanne Scheffler, Sian Tjia was successful in efficiently organising administration and teaching in our department.

Support for research in molecular virology at the Institute of Genetics, 1972–2004

Research in Molecular Biology requires research funds at a continuous and substantial level. Although the University of Cologne was in a position to grant us some highly appreciated resources, about 75% of the costs for personnel and more than 95% for consumable supplies and a lot of the instruments had to be procured from outside sources. Over the period 1972 to 2002, the research funds for the Department of Medical Genetics and Virology came from the following agencies: Deutsche Forschungsgemeinschaft (DFG, SFB74 between 1972 and 1988, and SFB274 from 1988 to 2000. For both programmes, I served as the *Sprecher* between 1978 and 2000); Genzentrum Köln; the Center for Molecular Medicine Cologne (CMMC); the Ministerium für Wissenschaft und Forschung des Landes Nordrhein-Westfalen (NRW); the Bayerische Ministerium für Landschaftsgestaltung und Umweltschutz; the European Union; the Alexander von Humboldt Foundation; the Fritz Thyssen Stiftung; the Wilhelm Sander Foundation; the Fonds der Chemischen Industrie; and amaxa GmbH in Cologne.

Sabbaticals and extended visits to other research groups abroad

Conducting research with a larger group leaves the principal investigator little time to do laboratory work himself. I therefore very much appreciated the opportunity provided by the University of Cologne for sabbatical semesters. Here is a list of my sabbaticals between 1978 and the present:

Stanford: 1978 (Paul Berg), 1993 (Uta Francke), Princeton: 1986 and 1999 (Tomas Shenk); Vanderbilt: 2006 (Ellen Fanning); and the list of extended laboratory visits: Rudbecklaboratoriet, Uppsala Universitet 2002, 2006 (Ulf Pettersson); Akademia Nauk, Moscow, Russia 1990 (Georgii Georgiev); Kawasaki Medical School, Kurashiki, Japan 1988, 1995, 1998 (Keiichi Hosokawa, Yoshihito Yawata).

Transgressing the Iron Curtain

One of the very interesting scientific contacts, starting in the mid-1975 and continuing into the 1990s, developed out of an initiative taken by Hans-Georg Zachau in Munich and Alexander A. Bayev at the Institute of Molecular Biology of the Academy of Science in Moscow. Over many years, we continued to have stimulating joint scientific meetings with our Russian colleagues, even at a time when travel between the East and the West was not commonplace. It was indeed a great experience to get to know leading scientists in Moscow: Akademik Engelhardt, Georgii Georgiev, Tatjana Venkstern, Andreji Mirzabekov, Lev Kiselev and many younger Russian scientists. Apart from the insights into the scientific accomplishments of a highly competitive group of researchers in Russia, we also witnessed first hand the limitations that a closed society and the aftermath of the Lysenko disaster had wrought on our Russian colleagues. Due to the immense kindness and sincere interest of Tatjana Venkstern, who had spent some of her high school years in the 1920s in Berlin, we also became familiar with, and certainly most favourably exposed to, the very rich cultural life in Moscow and other Russian cities. What important lessons we all learned during these exciting visits! I wish to add a special thank you and appreciation to Hans Zachau, Tatjana Venkstern, Alexander A. Bayev, Andrei Mirzabekov, who died much too early, Georgii Georgiev, Horst Feldmann and, of course, to my high school teacher in Russian, Professor Edmund Sandbach from Prague.

The Cologne Spring Meetings

These meetings were initiated by the Institute in the early 1970s. The faculty of the Institute took turns in organising the meeting. Thus the topics changed according to the scientific interests of the organiser(s). The meetings were supported by the DFG, the Fonds der Chemischen Industrie and private contributors and enabled researchers at the Institute to invite leading scientists in their fields from around the world. The Cologne Spring Meetings to this day — the tradition is successfully continued by the present faculty — have been a major attraction to many young scientists and students, not only in Germany but from many neighbouring countries as well. We all value them also as a possibility for a homecoming to the Institute. Our group organised or helped organise about ten of these meetings on topics of interest to us. In 1981, we invited our colleagues from around the world working on the then new topic to the First International Symposium on DNA Methylation.

Temptations to leave the Institute of Genetics

My former mentor at Rockefeller University, Igor Tamm, reminded me that every laboratory move to a different location will most likely be associated with a loss in scientific productivity. Nevertheless, I pursued with interest three attractive offers I received from the Universities in Innsbruck in 1981, in Munich in 1983, and from the Freie Universität Berlin jointly with the Schering Company to set up a new Institute of Gene Biology in Berlin in 1985. Each offer had tremendous allure. I managed to resist and, although acceptance of any of these offers would not have been a mistake at all, I never regretted my decisions. Turning down an offer by another university, always means that colleagues at these places had to be disappointed. To Hans Zachau in Munich, I owe a special thank you and expression of regret. I am also reminded of the statement by Heinz Huebner, one of my respected older colleagues from the Faculty of Law in Cologne: *"...und dann, Herr Doerfler, nur der erste Ruf ist schön."*

Die Emeritierungsurkunde

On 31 August 1998 I had the honour to spend about 30 minutes with the *Rektor* of the University of Cologne and to be handed the above. With enthusiasm and energy, I took this new opportunity for a fresh start and actively resumed many national and international collaborations. There were additional things to be appreciative of: Hartmut Thomas from the Ministerium für Wissenschaft und Forschung in Düsseldorf had offered a *Geschäftsbesorgungsvertrag* that allowed me to continue unabated research and teaching at the Institute of Genetics for another four years.

The reason I followed the very kind invitation by Bernhard Fleckenstein at the *Institut für Klinische und Molekulare Virologie* of Erlangen University to move our laboratory there in 2002 was not that I would not have obtained laboratory space in the new Institute of Genetics. For personal reasons, Helli and I opted to spend more of our time in Mittelfranken and some in Cologne. So, I enjoy the status of Professor emeritus, a life in transition, as the Latin verb *emereri* — *to* serve (deserve) out one's time — implies. It has been a pleasure and honour to have been granted 32 years in molecular medicine's service at the Institute of Genetics in Cologne.

In closing, I wish to thank my wife Helli and our son Markus for their tremendous support and unwavering understanding throughout the demanding years of a scientist's life. Without the unprecedented years of peace and prosperity in Germany and Europe between the 1950s and the present, we would not have had the chance to continue our work and maintain the many international contacts. I wish to devote these pages to the memory of my grandfather, Walter Kreuter, who in 1914 worked for a leading engineering company in Zurich, Switzerland and had offers to join the technical faculties in Prague and in Graz. With the collapse of normal life in Germany between 1914 and 1945, his opportunities were spoiled and never regained before his death in 1952.

20

T4 Hets and 5 Floors to Hang Around

Rainer Hertel

As a DFG-paid post-doc I spent three phage years (1963–66) at the Cologne Institute of Genetics. Before, I had done Ph.D. work in plant physiology, and after 1966 most of my research was again in plant science. During my doctoral auxin transport studies in the Horticulture Department at Purdue, I did a side job with phage T5 in the basement of Seymour Benzer's lab, where I met the visiting Carsten Bresch. Then I came to Cologne to study heterozygotes (hets) of T4 on the 5th floor, and as a side job, I was hanging around on all the other floors of the Institute. I published some papers and, in those good old days, Carsten's name was not on the title, although, of course, he was sponsoring, stimulating, criticising and helping.

My work was typical research on molecules without working with molecules: an indirect analysis with high-tech instrumentation, such as glass (chop) sticks, or toothpicks and agar plates. I looked for triple hets in my phage yield (see Fig. 26, top) after infecting bacteria with three different T4rII phages; the terminal redundancy and the two DNA-strands should provide four places for alleles.[78] Brute force, *viele Platten* and a little bit of thinking produced a few phage plaques, where the one starting phage particle must have carried three different, allelic markers.

---

[78]Rainer Hertel, "The occurrence of three allelic markers in one particle of phage T4", *Z Vererbungslehre 94* (1963), 436–441.

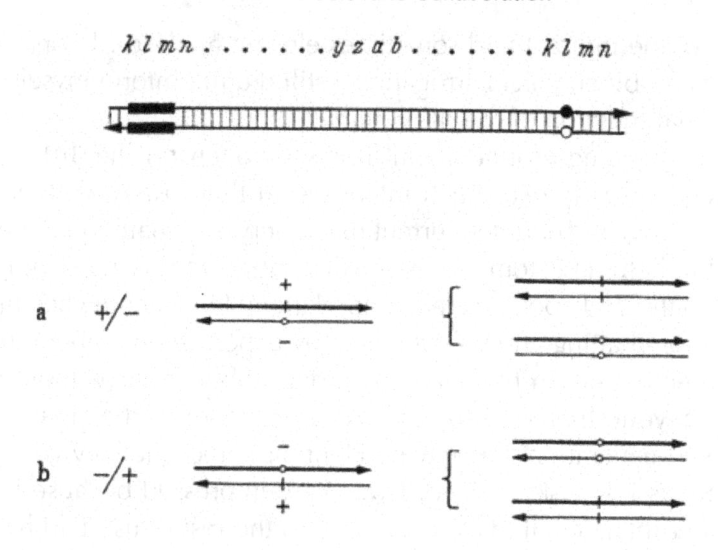

Fig. 26. T4 Hets. *Top*: T4 DNA, with terminal redundancy, carrying three alleles: wildtype (full circle), rIIC204 (circle) and rIIdelC33. *Bottom*: Heteroduplex and active (+) DNA-strand wildtype (|), rII227 (circle) (Hertel 1963, 1965).

But are these different strands also different for gene expression: one strand where the message is read off, and the other strand where the message is not read off? To solve this problem I looked for the survival of +/– heterozygotes (heteroduplices) in hosts where the function is needed before replication. The prediction was that half of the hets should die if there was only one active strand, and this was my result. In cases where you did not need the function before replication, you would not find the 50% loss, and this was my control. Again, a paper out of many agar plates and some thinking.[79]

Unfortunately, at the same time (or probably a little earlier), Julius Marmur *et al.* got a very simple biochemical answer for my question: the mRNA molecule hybridised with one strand of the phage DNA only.[80] (To be overtaken by straight biochemistry was common for

[79]Rainer Hertel, "Gene function of heterozygotes in phage T4", *Z. Vererbungslehre* **96** (1965), pp. 105–115.

[80]Marmur J, Greenspan CM, Palecek E *et al.*, "Specificity of the complementary RNA formed by Bacillus subtilis infected with bacteriophage SP8", *Cold Spring Harbor Symp. Quan.t Biol.* **28** (1963), pp. 191–199.

many of the indirect and sometimes elegant analyses. I confess to a rest of anti-biochemical arrogance: I still did not inform myself about the actual function of the rII-protein.)

I was hanging around on all floors — often on the 3rd floor — although I was not so much interested in their tRNA-work and not impressed by the counter-current machinery. My main job there consisted in helping Feldmann *et al* with writing songs and staging parties. On the 2nd floor, where our good friend Ulf Henning had his lab, I went for chatting and volunteering my expertise on amber mutants. I even ended up on two Henning-group publications without doing much beyond hanging around.[81] As a precursor of the great second Weyertal-generation, one day on Henning's floor the very young and slim Klaus Rajewsky arrived. I was very impressed because he had this finer intellectual make compared to the rest of us. (Did he wear a whiter lab coat?)

I enjoyed the science on the 1st floor. There was Peter Starlinger, *die reine Vernunft*, and, among others, the young, dynamic Heinz Saedler and Elke Jordan, and the mutants which nobody understood: somehow block mutations, but reverting, and certainly neither amber nor ochre. The 4th floor was photobiology, UV-inactivation and serious people: sometimes hard to understand, but always helpful. A fringe benefit of my running around was a detailed knowledge of where to find specific technical expertise and where to find bottles with useful chemicals (of course, always asking if I could take an aliquot). Thus our floor saved a lot of money, especially when Carsten had left for Dallas and we continued for some time as a *Jugend-forscht* group.

Names come to mind from the three years in Cologne: Ingeborg Schmidt (my TA), Jürgen Wiemann, Wolfgang Michalke and *Fräulein* Hildegard Ohligs (now Frau Michalke), Elke Rottländer, and so many

---

[81]Dennert G, Hertel R, Deppe G, Henning U., "Action of an amber suppressor gene carried by phage Phi80", *Z Vererbungslehre 97* (1965), pp. 243–254; Henning U, Dennert G, Hertel R, Shipp WS, "Translation of the structural gene of the *E. coli* pyruvate dehydrogenase complex", *Cold Spring Harbor Symp. Quant. Biol.* **31** (1966), pp. 227–234.

others. Rereading the papers, I got stuck in the acknowledgements, nostalgically, and I would like to add Anita Hoffmann from the DFG, who does not appear in our lists, although she really contributed a lot of support, moral support, too.

I was happy and content with these five floors, and one might think it is a pure blessing to have such a beautiful atmosphere. But now I realise that it was also socio-scientifically dangerous, because in these three years I never set my foot into the Zoology Institute; I did not even know where it was. And I did not go to Biochemistry or to the Botany Institute next door, although I was a botanist. I went once, only once, to *Entwicklungsphysiologie,* to Frau Harte, following Carsten's advice: "Go to Frau Harte, and ask whether your triple hets are statistically significant." There is a real danger if you are part of an active, nice family: you do not go outside of the family. This point was discussed at the workshop after Karl Wolfgang Mundry's talk about tobacco mosaic virus. It is paradoxical, that in my botanist mind, there was TMV, and there were the phages, but they did not connect; correlations were not activated between the Cologne phages and the TMV in Tübingen.

Soul searching, you find out later that you had been living unconsciously with antagonistic drives or with inconsistent motives: to mention the two aspects of science, the applied research to help in fighting famine, cancer, to help the Defence Department; one could call that aim *Weltverfügung* (manageability of the world), and on the other side, there is *Weltvertrautheit* (familiarity with the world). I must say, here in Cologne, the aspect of *Weltverfügung* never crossed my mind. I do not remember that I ever thought of the importance of our research for applied areas such as medicine.

An intrinsic contradiction here in Cologne was our feeling that we were anti-establishment, and after a while, of course, we became establishment. You may hate it, but you depend on establishment's acknowledgement to build your own position.

Carsten Bresch told me to do some serious teaching, "not just our phage course". Thus I was assigned to assist Professor Karl Esser in a genetics lab course with *Drosophila*. This was humiliating. I was studying all these beautiful, fashionable phages, and now I had to grow old-fashioned *Drosophila* flies and maggots on stinking maize polenta, spiced up with Nipagin. I had to fight with moulds and mites, and not with ideas. And the mites could only be removed by serial dilutions. For sure, this *Drosophila* would never come back to research! Of course, I was wrong, and now, after this lesson, I would advise young students to pick up an old-fashioned, leftover problem, and not to do fashionable things that "everybody" does.

Finally, a very sad story. We had a young researcher here who suffered from serious mental illness. I did not understand the seriousness of the case, and tried to help in some way. He called me in the middle of the night saying that his phage strains had been stolen from the refrigerator and his data book, too. I should come and see, he said; I drove in, and nothing was stolen. Then a visiting friend of mine was identified by our sick colleague as a policeman from Erlangen, sent by the CIA and/or the KGB, etc. This was scary enough; but the scariest thing was that this man's antibodies, his experimental data, were perfectly fine. His experiments took him longer than other people but they were correct. I take it as an allegory: we scientists are nuts. Nevertheless, sometimes, somehow, something true and solid results from our work.

21

The Long Way from Glucose Effects in Bacteria to (Systems-) Biology

Joseph W. Lengeler

Looking back over 50 years of the Institute of Genetics at Cologne, I shall present the view of a first generation Ph.D. student who started his studies under the "old" German university system, with its traditional structures and professors, and was thrown overnight into the "new", American-style, system. Furthermore, I will show how particularly this training influenced my later scientific career, and indicate how the institute influenced genetics and biology in Germany.

1956: "Traditional" biology teaching in Cologne

I began my studies at the University of Cologne in 1956 as a foreigner, having obtained the *Abitur* after a six year curriculum at the "Bischöflische Schule" in St. Vith, Belgium. Although this had been an excellent college, I had begun to dislike heartily the Belgian and French school systems. Admittedly, they were fast, but retained even at the university level strictly regulated curricula; obviously predecessors of our actual speedy bachelor and masters curricula. In contrast, the German university system was very flexible and required mostly independence from its students. Here, a cocktail of lectures, seminars and laboratory courses was offered from which every student had to choose for themselves what suited best. This freedom allowed me, as a Belgian citizen, to begin studies in biology, chemistry and whatever else I wished.

For an inexperienced student, university professors seemed to be "beings from another world" which one had to approach with due respect, never exceeding a certain safety distance. Fortunately, after some laboratory courses, excursions and, in particular, a few interesting Karneval-parties, most professors and teachers became almost normal human beings, with the exception of only a few that were neither respected nor liked. Training in the laboratory was mostly through post-doctoral fellows and other students. Presentations in public of the results from thesis work or the data from other groups, followed by a serious discussion, were rare, and the contributions by most "established" scientists in seminars were not very revealing. In contrast to physics and chemistry, the technical equipment of most laboratories in biology was very limited, mostly due to the lack of funding for their "non-application oriented" research, and the methods used were often not state-of-the-art. The favoured scientific concepts were rather "traditional", i.e., oriented towards a qualitative description of biological structures, physiological phenomena and organismic diversity.

Differences between the German and the American universities and scientific style

In retrospect, I see three important differences between the American and German university structure and style in biological research which, in an alleviated form, still exist. One must be aware of these if we want to understand the impact of the Institute of Genetics on genetics and biology in post-war Germany.

First, traditional secondary school training in Germany covered a broad range of topics, from the humanities, foreign languages and universal history, to mathematics and natural sciences. Hence, a comprehensive view and broad topics in scientific research were favoured, e.g., problems relating to evolution, developmental biology, brain research and physiology. Such long-lasting projects required a solid knowledge in the literature and much reading, a trend reinforced by the temporary ban of German scientists from top international meetings. Americans, in contrast, still have a less comprehensive college education, and scientists prefer a practical approach, asking limited questions that produce "more results more quickly". Also, the development of rapid assay techniques for immediate application, e.g. in industry, is typical, as well as the relegation of very complex problems to later. Finally, instead of reading, Americans prefer to participate regularly in high-level meetings to learn the latest developments instantly.

Second, in contrast to physics and chemistry, applied biology in the form of medicine, biotechnology, or plant and animal breeding had been separated for generations from "pure" research, leading to a certain "aversion to industry" among academic biologists. This trend was intensified after the infamous "eugenics" and "euthanasia" practiced under the Nazi regime. Applied research was consequently delegated preferentially into clinical departments, Technical Universities and non-university institutions like the Kaiser Wilhelm Gesellschaft and its successor, the Max Planck Gesellschaft.

Third, structures and assignments of the German and American academic institutions differ. Thus, German scientists paid by the university have to devote half of their time to research and half to teaching and administrative tasks. Although, in principle, the topics in teaching are free, many curricula that lead to official state positions, e.g., college teachers, physicians, lawyers, etc., are heavily regulated by the state. Furthermore, the staff-to-students ratio is very low, the supply in staff and finances inadequate. Particularly important for top American research are graduate schools in private universities, blessed with comparatively generous endowments and funds for research, and a light teaching load. Their need for external funding, and hence service orientation, requires a powerful and independent administration with the power to react quickly by closing outdated institutions, and hiring young promising scientists by means of attracting individually allotted salaries. This allows for a qualitative growth of departments by diversification, and the system encourages connections to industry and application.

In Germany, basically no university is private, and only a few non-university institutions, in particular the Max Planck Institutes, have relative freedom and adequate facilities. The university system was and is centred on a powerful *Ordinarius* (full professor), more often than not the only representative for his area of research at his university. A way of providing for both the rapid growth in student numbers and new fields in the sixties — obviously needed after the two world wars with their heavy brain drain and death toll — was to increase in a quantitative way the groups of such full professors. Unfortunately, very often the price paid for this cheap solution was scientific immobility, an insufficient number of positions, inadequate salaries for young scientists, and a devastating teaching load for all.

Joseph Straub, a "modern" biologist in Cologne

Fortunately, the supervisor of my final thesis work was Joseph Straub who has already been mentioned repeatedly during this symposium. Straub was one of the respected professors whom, over the years, I increasingly learned to estimate and to like. He was by training a botanist, with a strong interest in cytogenetics, and in particular the genetics of incompatibility in plants and the role of chemotropism between the egg cells and the pollen tubes. During my thesis work, I had succeeded in showing unequivocally the chemotaxis between the egg cells and the male spermatozoids, as well as the existence of species-specific "gamones" or sexual hormones in the moss *Sphaerocarpos donnellii* and its relatives. Consequently, Straub offered me the opportunity to start as a Ph.D. student in his group, and to isolate as well as to characterise biochemically these gamones, which was quite an honour for me. Unfortunately for him, I had in the meantime heard lectures and participated in laboratory courses given by C. Bresch, W. Harm, P. Starlinger, J. Raper, a well known expert in fungal genetics from the United States, and other guests of his institute. These guests had been offered shelter in his institute and had been asked in return to help in modernising our training in botany, genetics, and biochemistry.

Straub was deeply convinced that there was a new and fascinating genetics which, unlike Mendelian genetics, did not confine itself to the simple description of phenotypes and genes. It would analyse the underlying molecules and the biochemical mechanisms involved in gene expression, and a similar combination of genetic and biochemical methods should also be introduced into modern botany. However, these ideas prompted me to leave his institute in order to get such modern training in the neighbouring institute. Straub had no apparent objections and we remained to his death on very friendly terms, another example of his unselfishness and generosity.

The start in the Institute of Genetics

In August 1962 I started my Ph.D. thesis with P. Starlinger in the illustrious Institute of Genetics, with the "famous" Max Delbrück as its director. My first contact with "Max" had been while I was still in Straub's group. One day in 1961, I had to fetch the key for the greenhouses from the secretary. She was on the phone ending her conversation by a polite "*Jawohl, Herr Minister*!" laid the receiver down and went into Straub's office. Through the open door I could see Straub and a somewhat shabby person, lolled in an armchair with his legs hanging over an armrest. The unknown person said after a short while: "You tell Herr Minister, either he has time for me today, or he can visit me from the day after tomorrow on in Los Angeles". Back at the phone, the secretary gave this answer, and to my total amazement, the person got the appointment for the same afternoon. This, as I learned later, had been Max at his best!

In 1959, Jacob and Monod had published their famous operon model. It was based mostly on mutants, cis-trans dominance tests, mapping by means of conjugations and transductions, and some simple enzyme tests. This was a brilliant example of what excellent genetics can achieve. However, most traditional biochemists could not be convinced by such arguments, in part due to the "dirty biochemistry" used, but mostly because they simply could not understand genetic arguments. Worse still, traditional geneticists could also not be convinced since "everyone knows that bacteria have no nuclei, no chromosomes, and hence no decent genetics". Despite such objections, Starlinger's group had begun to analyse, similarly to Jacob and Monod, a more complex gene cluster, the *gal*-regulon for D-galactose uptake and metabolism in *Escherichia coli* K-12. Regulation of this relatively complex regulon included, besides a repressor GalR, its repression in the presence of D-glucose. This phenomenon, originally known as the "glucose effect", had been renamed in 1961 as "catabolite repression" by B. Magasanik (Cambridge, USA), and in analogy to the LacI-repressor for the *lac*-operon, and the cI-repressor of bacteriophage lambda, was thought to involve a "catabolite- repressor".

Thus Starlinger proposed to isolate mutants with defects in this repressor and to map the corresponding gene. Fortunately, no one, and certainly not me, knew that mRNA and promoters, transcription and translation, as well as the difference between positive and negative control had still to be discovered; how complex gene regulation would quickly become; that carbohydrate transport and cell membranes, one of the worst terra-incognita in biology of the time, were central for my problem; and that mapping of a mutation on a primitive gene map with less than 100 markers only by means of conjugation and P1-transduction, could become a major endeavour.

Immediately, a few obvious differences between the style of working in the new institute compared to that in traditional groups became clear. First, there was not just one, but five group leaders, and all five had de facto the same status, regardless of their official position. This contrasted with the rest of the biology departments, which also had five professors, but in total. Since the teaching load was comparatively low and, due to Max's reputation, the funding, according to German standards, very good, the Institute had a status comparable to a Max Planck Institute. All doors in the Institute were always open, and their backside invariably was a blackboard. After lunch, we regularly had coffee, which Starlinger generously paid, and discussed our data or politics, hence the door-blackboards. We were encouraged, and quickly learned, to present and discuss our data in more or less elegant English before any of the numerous guests, who ranged from post-docs and professors to Nobel Prize winners.

There was a weekly journal club, obligatory for everyone from Max to the last student, in which interesting papers, but also reports on international meetings, were presented and seriously discussed. Regular presentation of the progress in our work in front of the entire institute was expected, and, if insufficient, had to be repeated. We were also encouraged to teach the new genetics to, for example, elder biology teachers in colleges, or to discuss it in public. My first, unforgettable such lecture was 1963 in front of 100 Catholic nuns during their training as clinical nurses. The *directrice* had learned her genetics under the Nazi-regime, and had subsequently refused to teach such "racism and eugenics". Because, however, exams in genetics were obligatory for nurse students, she now wanted to learn for herself what modern genetics was all about. I was immensely proud when after two days of lecturing I had convinced her and her students of the fascination and importance of genetics for their future work. Henceforth, modern genetics was included in their curriculum.

Through such innovations, our knowledge both at the theoretical and at the methodological level increased rapidly. In parallel, our independence and self-confidence also grew — sometimes approaching arrogance, according to students from other institutes. Within a year or so, I had isolated several mutants with an altered catabolite repression for the *lac*-operon and the *gal*-regulon. I had even mapped the corresponding mutations, but, totally unexpectedly, in the *lacI*-gene and not in a "catabolite repressor" gene.

During the next three years I learned the hard way, first, that "any fool can isolate mutants, but it takes a man to understand them", and second, that not every hypothesis published by internationally renowned experts needs to be correct. Indeed, a closer analysis of my mutants showed that a major part of the "glucose effects" was due to the fact that in the presence of glucose the uptake of less efficient carbon sources like lactose and D-galactose into the cell, and consequently their role as inducer, was inhibited, and that *lacI*-mutations overcame this inhibition. And thorough literature research showed that this "inducer exclusion" had been described already in

1942 and explained correctly by J. Monod in his Ph.D. thesis, but was, curiously, overlooked or dismissed by Magasanik as less relevant. Consequently, the major conclusion of my thesis, and of the thesis of S. Adhya at Stanford, published simultaneously in 1966, was that the glucose effects in fact are a combination of inducer exclusion and of catabolite repression.

At about the same time, Heinz Saedler had discovered during his thesis work in Starlinger's group a class of "strongly polar" mutations in the *galKTEM*-operon with the curious property that they reacted to every mutagen. To help a friend working in the institute of botany on extra-chromosomal inheritance in *Oenothera*, another speciality of "German" genetics, we had agreed to organise a special seminar on non-Mendelian genetics. Although Max had always claimed that he did not understand the papers of B. McClintock, but he knew them to be very important, Heinz was brave enough to give a seminar on the various "mobile genetic elements" in plants and animals. This seminar obviously accelerated the recognition of such "insertion elements" in bacteria by Saedler, Jordan, and Starlinger, one of the first internationally outstanding results from the new institute.

The transition period from Max's departure to *Sonderforschungsbereich* 74

Already in 1965, Delbrück, Bresch, and Harm, together with most of their co-workers, had left the institute. This left Starlinger and his group the task to reorganise the institute, and in particular to do all the teaching. Fortunately, U. Henning, H.G. Zachau, B. Müller-Hill, and later W. Vielmetter, P. Overath and K. Rajewsky could be hired as new group leaders. Their arrival quickly shifted research at the institute away from bacteriophages, photobiology, and the physicochemical properties of DNA — first towards molecular genetics, bacterial physiology that included transport systems and membranes, RNA-sequencing, and the increased use of biochemical methods; and later towards immunology and virology. As a direct consequence, the traditional "spring course in phage genetics" became the "spring course in phage and bacterial genetics", and gradually shifted also towards the new topics.

From the start, the spring courses, for several reasons, had an immense effect on the rapid spreading of the up-and-coming molecular biology in Europe. First, because the latest methods were taught; second, because the participants came from all over Europe and ranged from students to prominent professors; and third, because the courses were completed by a meeting, in which world famous experts participated, including Max himself. The meeting ended traditionally, in Max's honour, with one of his beloved parties.

In 1966, I was hired by P. Starlinger as a post-doc and began to shift my research to the transport system for D-galactose present in *E. coli* and, perhaps not surprisingly, back to chemotaxis. The first topic seemed to be a logical continuation of my thesis work, while the latter topic was triggered by a paper from J. Adler (Madison). Adler had shown that genetic methods could also be used to analyse chemotaxis in this bacterium. I speculated that *E. coli,* with its limited gene pool, might use parts of its highly specific transport systems as stimulus receptors, and direct tests of our transport mutants quickly confirmed this idea. Together with two physicists, K.O. Hermann and H.J. Unsöld, for whom pure physics had become too boring, these projects were tackled and became the basis for a joint project in the famous *Sonderforschungsbereich* 74 that, among the first of its kind in Germany, started in 1969.

Encouraged by the work of P. Overath's group on phospholipids and biological membranes, we intended to purify the galactose transporter, to insert it into artificial, so-called black-lipid membranes, and to analyse its properties. In retrospect, it seems clear why this ambitious project was doomed to fail. *E. coli* has seven transporters for D-galactose that comprise ion symporters, ABC-transporters, and phosphotransferase (PTS)-transporters. Furthermore, biochemists did not easily accept Mitchell's "curious" ideas on the role of ion gradients in active transport, and yet the lactose-"permease" LacY was the only accepted paradigm of any decent transport system.

From Cologne to Harvard Medical School and back to Germany

As a consequence of this highly exciting but controversial period in my scientific career, I decided to leave Cologne, mostly to get a better training in membrane biochemistry at a top American university. During one of his frequent visits, Max tried to convince me to enter sensory biology, his latest area of interest, and to join Adler's group. But in the fall of 1969 I started instead at Harvard Medical School in Boston, where I joined the group of E.C.C. Lin in the department of bacteriology and immunology. Ed Lin's group had been working, among other projects, on catabolite repression, phosphotransferase systems, and carbohydrate transport, using mostly biochemical methods. Very soon it became apparent that in compensation for my — by the way excellent — training in biochemical methods, I was expected to introduce genetic methods into his projects. This was quite an experience in a department famous for its geneticists, e.g., J. Beckwith, B. Davies, D. Fraenkel, H. Kalckar and L. Gorini. Success in this endeavour was an excellent confirmation of the high quality and standard of our training in Cologne. However, to my great disappointment, I learned that Kalckar and J. Adler also had had the clever idea of carbohydrate systems being used as receptors in bacterial chemotaxis, and had already submitted it in a, by now famous, paper to PNAS.

And thus I chose, in the beginning somewhat grudgingly, as the topic for my own research the strange "transport systems" called phospho*enol*pyruvate (PEP)-dependent carbohydrate: phosphotransferase systems, or PTSs. Over the next 30 years, however, this choice turned out to be an excellent one, mostly because PTSs have central functions in carbohydrate transport, chemosensing and global control, and because their modular evolution is especially clear. Thus, I worked, funded in three additional SFBs, on each of these topics, first at the University of Regensburg, and later at the University of Osnabrück.

PTSs constitute, first, a series of carbohydrate transport systems of which up to 20 different types can be found in a single bacterium. All

PTSs of a cell are coupled to a single phosphotransferase complex involved in the vectorial phosphorylation of the substrates during uptake. We identified, isolated, and characterised in several strains of enteric bacteria all components from 12 PTSs, some to the crystal level. Because each transporter generates intracellular substrate-phosphates, we also identified 10 related catabolic pathways. Based on sequence comparisons, many orthologous and paralogous genes could be identified; the convergent evolution of several metabolic pathways, including several for non-PTS carbohydrates, could be retraced, and the modular evolution of the various components could be mimicked, e.g. by creating active heterologous complexes. During the analysis of a set of sucrose-specific PTSs from enterobacteria, the first large conjugative transposon (100kb) was found and shown to comprise elements of bacteriophages and of conjugative plasmids, and to be closely related to the medically important pathogenicity islands.

Very early on I had discovered that each PTS-transporter was simultaneously the chemotactic receptor for its substrates, and postulated an ATP-dependent process that should couple them to the rest of the chemotactic machinery. This hypothesis was brilliantly confirmed when several groups proved the central role of an ATP-dependent two-component system in chemotactic signalling. We were later able to show that and how a PTS-specific protein-kinase modulates this signalling, and hence, chemotaxis towards carbohydrates.

Over the years, many groups, including ours, were able to show that the fast control of the activity of non-PTS-carbohydrate transporters, i.e. inducer exclusion, depended on free IIA^{Crr}, one of the components of the PTS. Catabolite repression, in contrast, was due to transcription control of the large group of genes involved in catabolism and energy generation. The alarmone or second-messenger cAMP, together with a cAMP-receptor protein CRP functioned as a global activator for these genes. Again, IIA^{Crr} was involved in this control since in starved cells: IIA^{Crr} was phosphorylated, activated the adenylate cyclase Cya, and hence allowed gene activation by the cAMP-CRP complex. In its dephosphorylated form, i.e., when the cell was well supplied with carbon sources, IIA^{Crr} was inactive.

Strictly speaking, catabolite repression thus corresponds to the lack of gene activation through cAMP-CRP. After it had become clear that the phosphorylation state of the PTS-component IIA^{Crr} was crucial in inducer exclusion and in catabolite repression, and that the PTS-specific protein kinase was involved in chemotaxis as well, I began to realise that all components of the various PTSs from a cell form a complex signal transduction system. This system is used by the cell to distinguish between feast and famine, and to modulate accordingly, by a global and epistatic regulatory network, the transcription of about four hundred genes (forming the *crp*-modulon), and enzymes involved in carbon catabolism, energy generation, and the quest of food, including chemotaxis. No wonder in retrospect, that I did not understand my first mutants!

Conclusion: back from molecular physiology to (systems-) biology

From about 1985 on, I began to wonder how this increasingly complex system "quest for food" could possibly be described, first in a qualitative way through the emerging techniques of genomics and proteomics, and perhaps later in a quantitative way that should include the dynamics of the system, i.e. the various metabolic and sensory fluxes as they occur in an intact, living cell. Fortunately, in 1989 I met an engineer and expert in systems and control theory, E.D. Gilles, then at the University of Stuttgart, and now at the MPI für Dynamik komplexer technischer Systeme in Magdeburg. We began our fascinating cooperation by learning each other's scientific language and way of thinking. This was followed by a period in which we learned how to structure complex biological systems based on genetic units, e.g. regulons, modulons, and even larger genetic units, on global regulators, and the corresponding signal transduction systems. Together with the metabolic networks these elements form cellular functional units (CFUs), or *systems*, of ever increasing complexity at each new level of complexity. Next, we learned how to convert such structured biological systems into structured mathematical and computer models, and to test predictions and simulations originating from these models in continuous fermenter studies. Progress of this very successful strategy is hampered at present mostly by the lack of quantitative data from intact cells that allow the reconstruction of even fast changes in the flux of metabolic pathways and of signal transduction networks.

And thus, my retrospective of 50 years of the Institute of Genetics ends by the surprising conclusion that we have finally come close to the dream of the most progressive biologists of the end of the 19th century. To them, the ultimate goal of a biologist was a comprehensive understanding of the dynamics of complex biological systems that had to be based on a solid knowledge of the biochemical details. My training at the Institute had emphasised open-mindedness and enthusiasm for the various areas of biology, and taught me eventually both the German and the American style of solid scientific work. Clearly, these were major reasons why my thirty years as a biologist have not been unsuccessful, and why they have been great fun!

Discussion

Ute Deichmann

After you learned molecular genetics in Cologne you decided to go to Harvard or somewhere in the United States to learn more biochemistry. I find this quite unusual. Can you tell us more about this and also about the decision to go to the United States and not to Tübingen?

Joseph Lengeler

I wanted to go to the United States because I was told what a fascinating country it is, and because my wife, who for a while had been with Bresch in Dallas, wanted to go back once more to see other parts of the country. Against such strong arguments, there is no arguing! But seriously: from frequent discussions with our trained biochemists

B. Müller-Hill and P. Overath, I had learned how important biochemical methods had become in modern genetics, in particular for my research on carbohydrate transporters and membrane-bound sensors — those relating to membranes, integral membrane-proteins and phospholipids. Consequently, I got in contact with one of the world-leading experts in this area, Prof. E. Kennedy at Harvard Medical School, who, having himself no vacancy, suggested Ed Lin as an equivalent alternative. And he was right! I couldn't have found a better teacher, tutor, and later, friend than Ed. Excellent though Tübingen and its biochemistry were at the time, they could not have rivalled the membrane biochemistry group at HMS, nor Harvard with its multiple centres of excellence, nor Boston and its surroundings.

22

Attempts to Transfer Lab Structure and Scientific Habits from Harvard to Cologne

Benno Müller-Hill

I got an education as a chemist and when it came time to do my Ph.D. work, I decided that inorganic and organic chemistry were boring, but biochemistry was what I wanted. I did my work with Kurt Wallenfels in Freiburg, who was half organic chemist and half biochemist. I learned biochemistry there in his lab. When this came to an end I thought, "this is nothing for me, what I really want to do is to know how to get mutants." Because if you have a mutant you have a new substance you do not need to isolate again and again. When I had the chance to go to Bloomington, Indiana in the United States, to Howard Rickenberg, I accepted this opportunity, and started to learn how to work with *E. coli* and bacterial genetics. From there I went to Harvard, to the Watson-Gilbert group, and worked with Wally Gilbert.

We had just one goal, the isolation of lac repressor. The first thing I did there was to isolate two nonsense mutants in the lacI-gene which would prove that the lac repressor is a protein and not, as Jacques Monod had published, an RNA. To do this experiment was Wally's idea, but I did all the practical work and when this was published Wally was not co-author. He had the belief that he was only the co-author of a paper if he contributed a lot to it — at least 20% of the practical work. And in that particular case he had not done anything, so I was the only author. After that I isolated a mutant that bound the inducer of lac repressor better than the wild-type and Wally used it successfully for an *in vitro* test of lac repressor. In 1966 we published a paper on the isolation of lac repressor and a little bit later I got some wonderful mutants which overproduced lac repressor. Now you could produce as much lac repressor as you wanted. Lac repressor had become an ordinary protein.

In 1967 I got a letter from Peter Starlinger offering me a full professorship at the Institute of Genetics here in Cologne. When I got this letter I went to Jim and asked him: "Look here, I got this letter, what shall I do?" He said: "You have to accept it!" I sat down and wrote a two-line letter saying that I accepted this kind offer. And, of course, if you consider this today it was total stupidity to do this. There was no deal. I didn't ask how many collaborators or how much money I would get, just nothing. I was totally *weltfremd* to put it mildly. In April 1968, in the beginning of the *Sommersemester,* I arrived in Cologne by plane at 8am, and the same day at 5pm in the afternoon I gave my first lecture on bacterial genetics. Again I must say that I had never heard a lecture on bacterial genetics and I had never given one myself. It was my first lecture.

So life started in Cologne and this was fantastic because the Institute was great. I wanted to structure my lab similarly to what I had experienced in Wally Gilbert's and Jim Watson's lab. The first thing was that nobody had to wear white coats in Harvard. In German chemical institutes usually everybody was wearing lab coats. When I came to Cologne nobody was wearing lab coats anymore in the genetics institute. When I was invited to give a talk I usually wore blue jeans. Blue jeans were considered to be against all cultural norms here in this country. Today this has changed. Even here in Germany, people are now wearing blue jeans.

The second habit from Cambridge I wanted to continue was that one is only co-author of a paper if one had done some practical work on it. And I give you two examples of my lab experience. I had a post-doc, Konrad Beyreuther, and three graduate students who were working with Konrad. Konrad had the goal to sequence the protein lac repressor. For this purpose we isolated 11 grammes of lac repressor with my good mutant. Konrad organised the sequencing. For this he needed about 10 grammes. When he finished the sequence it was clear that I should not be co-author. But at that time (it changed, of course) Konrad had a real problem with writing in English, so I sat down and wrote the paper for him. My name just appears in the acknowledgement. My second example is that I had a post-doc from the United States, Jeffrey Miller, who was very smart. Jeffrey had the idea, together with Geoffrey Zubay, to isolate Trp-repressor. In order to do so he collaborated with a student of mine and they isolated Trp-repressor. They wrote a PNAS-paper and again I was not co-author.

This went on for almost 10 years, and then came an event that led to the end of this strategy. I had a post-doc, Bruno Gronenborn, who collaborated with Joachim Messing, now a well known geneticist. Back then he was a post-doc of Peter-Hans Hofschneider in Munich. Bruno and Joachim tried to convert the phage M13 into a vector. What they were doing was to insert the lac promoter, the lac operator and the beginning of the lacZ gene into M13 DNA. When they succeeded, the question turned up of who would be co-author? I thought of course that these two post-docs would be, but Peter-Hans naturally thought he would be co-author too because this was standard in Germany. Now I had the problem that if I would not be co-author on this paper and Hofschneider was, obviously the whole thing would be lost for me. I had done no experiments, but I spent some time discussing the experiments with Bruno. This was the moment when I gave up. I capitulated and became co-author. If I think back there are two ways of looking at things: one is that my policy was completely idiotic and just totally wrong; but on the other hand it worked tremendously well because students and post-docs got inspired by their tremendous independence. But the strategy is only possible if a majority of groups do this. Otherwise you are lost. If I could convince a majority of the people now working in the

Institute to do it that way I would be happy, but this seems to be impossible.

So much for science. Now I may just say a few words on a different topic. In 1978 I discovered by accident that the history of human genetics in Germany was not written. I discovered that there was a past which was incredible and that there was almost total silence about it. So I spent about half a year to read the relevant material and wrote a book on that. When the book, *Tödliche Wissenschaft*, came out in 1984 it had its first review in *Nature*. It was eventually translated into eight languages. It was also the starting point of a process which ended with the Max Planck Society creating a committee to investigate its own past. They finally accepted Mengele as part of their history and invited the twins who had been mistreated by him. Hubert Markl, the president of the Max Planck Society, apologised to them.

If I think back, it did not help my career in Germany to write this book. This is true. But I would do it again. I would do it again despite the fact that within Germany this was bad for me. I will give you some evidence for that here. During the 37 years I was here in Cologne I was asked only twice to write a *Gutachten* for a *Berufung* to a German university. I was never asked when one of my own students applied for a professorship. I received no prize, no honour in Germany. You see here somebody who is non-existent or almost non-existent. And again, if I would be asked whether I would do it again, I would say yes. After this was written I became honorary member of the German Genetics Society.

Discussion

Peter Starlinger

I have to tell an anecdote. Those of you who know Benno know how gentleman-like he is. He always wore a white shirt and a tie. But now came the time when he had to give his lecture at the Spring Meeting, so what did he do? He took off his coat and his vest and his tie and he unbuttoned his shirt, went to the lecture hall, gave the talk, and went back to the seat.

Walter Doerfler

I have to add something to Peter's story. You know molecular biologists have something of a frantic obsession of this neck tie thing — sometimes a little ridiculous, but nevertheless. Max Delbrück had a perfect solution: when I saw him in Cologne he was like that, when I met him in Pasadena he had a necktie. When he came once on a visit here and had to give a talk he had a necktie but the necktie was underneath the shirt. That was his Cologne modification for wearing a necktie.

Ute Deichmann

I wonder how far this kind of punishment went [for writing the book]. For example were you asked to write reviews on grant proposals by the DFG?

Benno Müller-Hill

That is interesting. I never had any serious difficulty to get money for research and I was often asked to write reviews on grant proposals. But I think this was a different world. These were different people who were the *Gutachter* here.

I just want to add one anecdote which I forgot to tell. Once, Wally was here in Cologne working in Klaus Rajewsky's group on some mouse experiments. He was staying for several months and he lived in our apartment. One time Max Delbrück came and so we had dinner together and Max was speaking German and Wally did not understand German. I said to him: "Max, we should speak English. Wally does not understand German." And Max said: *"Er lebt in Deutschland; da sollte er auch deutsch lernen."*

Walter Doerfler

Of course, I communicated with the outside world as we all did and I got a lot questions and comments on Benno's book. May I say now that I always defended it. But I must say that in the scientific community in Germany the majority, at least in our generation — I cannot talk about the previous ones — was fully acknowledging the value of your book. There was absolutely no question that younger people who were interested enough to read it were enthusiastic. Although the official world can be harsh and stupid, I think the real science world in Germany acknowledged your book.

Benno Müller-Hill

I may add one anecdote in this context. Two weeks ago I was in Berlin at the final meeting of the Max Planck committee which was investigating its past. During the discussion Hubert Markl who was then president of the Max Planck Society told the following anecdote. In 1984 he bought my book *Tödliche Wissenschaft*. He was so moved by it that he bought 30 copies and gave them away to his colleagues. But this was not the official world. This was kind of non-official world.

Jonathan Howard

It seems to me that all people who told us these stories have done very well in their careers. It seems that, at least in the golden days, apparently not contributing authorship was not so dangerous. Is that correct, or are there instances in your remembrance of careers which were harmed by not putting one's name on the papers of one's students?

Benno Müller-Hill

It's getting difficult to ask for money if you are not co-author of important papers.

Maria Leptin

Fritz Melchers introduced a sort of hybrid way of dealing with this. His graduate students have to publish their main paper by themselves. There were many papers which came out of the lab which we all had our names on, but every graduate student had to publish one paper on his own. For me this is a very good way of handling this. I wish I had the strength to do it; I wish we all had.

23

Joining the Institute of Genetics Early On as an Immunologist

Klaus Rajewsky

I would like to begin with a remark related to Benno's talk. During the entire time I spent in Cologne I was also never asked to evaluate a professorship candidate at any German university. I wonder whether this was the same for the other colleagues in the Institute. I think the reason was that in Germany the Genetics Institute in Cologne was considered not to be in the mainstream, but being somehow left and full of foreigners, and so we were not asked to do these things. I was asked to evaluate grants but never applicants for professorships. Since I moved to Harvard I get such requests all the time.

I joined the Institute of Genetics in Cologne as a true outsider. I was not a molecular biologist, but an immunologist, and I did cellular immunology and immunochemistry; I had just done some biochemistry in my PhD thesis work. At the Pasteur Institute, against the will of my boss, I had listened to the lectures in the Jacob and Monod lab, which were obviously fantastic, but I did not understand anything. I was just fascinated by the way these people talked and argued; but then there seemed to be no connection to my own field.

My joining the Institute was thus a sort of a miracle. At a meeting I met Ulf Henning, who was a new member of the Institute. He listened to me when I talked about antibodies against enzymes and he said: "Why don't you come and join my group? I will give you a lab and a technician and then you can just do what you want." I thought this was an extremely attractive offer. I thought it would be nice to see what these young scientists were doing in this new science. So I went to the Institute and I got a lab, 12 cages for rabbits in the basement, and a technician, and then I was allowed to do what I wanted to do. Much of the time I would listen to the guys around me, hoping that at some time I would be able to do something in immunology which would get close to all those fantastic things they were doing: analysing bacteria and bacteriophages by mutagenesis. That was my dream over the entire time I worked at the Institute — some forty years. And at the end the dream became reality.

The first thing we did was that we started to breed rabbits to find genetic traits in their immune response. It worked out wonderfully. Finally, we had around 120 rabbits some of which responded very well to antigen and others that did not. This difference was inherited as a single Mendelian trait. Here came my first collaboration with the molecular biologists in the Institute, namely with Carsten Bresch's group. They wanted to immunise some rabbits with a virus to make antibodies. Of course, we allowed this, but when they immunised my rabbits, the rabbits became sick. It turned out that their virus preparation contained rabbit smallpox virus.

That led to a crisis in the immunology department exactly at the time when the whole Institute was in a crisis; only Peter Starlinger was still there and Carsten Bresch had already left. In our immunology department all the animals died except one and I thought my career was basically finished, because I had nothing to work on any more. To save myself, I decided to go to Mill Hill in London to learn from Avrion Mitchison how to work with mice. When I came back to Cologne we started to work with mice rather than rabbits, and things worked out very well, including our interactions with the molecular biologists. In the end we could do with our mice what the young molecular biologists had been able to do in bacteria and bacteriophages at the time when I joined the Institute, namely to analyse biological processes by mutagenesis. And we could even introduce mutations in a targeted and conditional manner! This was a most satisfying experience.

I have made a summary slide of how the immunology department of the Institute grew over the first and second 15 years from the one lab and the 12 rabbit cages (see Abb. 1). As you can see, a large fraction of the post-docs who came to the immunology department were from abroad. Indeed, over the entire time that I worked in Cologne I got amazingly few applications from German post-docs, much fewer than I now get at Harvard. In contrast to Max Delbrück's German speaking rule, I introduced English as the working language in the lab from the very beginning. The reason for this was very specifically that I wanted to attract foreign post-docs, because I thought that it was of fundamental importance for the Institute to have an international atmosphere. It was the only institute in Germany at the time, where this was generally accepted. Indeed, a vote was taken that we would have all seminars at the Institute in English! That made it easy for foreign post-docs to come to Cologne, and we had lots of international collaborations arising from this, of course. That again was extremely satisfying and a special element in the life of this Institute.

Looking back, I was mainly persuaded to come to the place because of the fascinating new science that was done here, the international perspective, and because of the very special non-hierarchical, relaxed way in which things were done at the Institute. I was shocked by what Rainer Hertel said, namely that I always had a clean lab coat on in the early days. I cannot remember that at all. I thought that I was always running around like now. But apparently it took me more time than I then realised to adapt to the new environment and a new way of doing things. I am immensely grateful that I had the chance to work for almost 40 years in this wonderful place.

Discussion

Jonathan Howard

When you insisted on English and stressed the valuable international collaborations it sounds like you had already long since forgotten the idea that the Institute of Genetics was there for the rehabilitation of genetics here in Germany. That wasn't on your mind?

Klaus Rajewsky

Indeed not. I had been in Paris before, living in an international environment, and that is what I enjoyed. Introducing molecular biology in Germany was of course initially a mission of the Institute, as the main place where molecular biology was done in the country. But I came there to learn molecular biology in the international context in which it was done, and when I began to build the immunology department, I wanted to preserve that very special flavour which everybody in the country knew, sensed and was somehow interested in.

Jonathan Howard

I mean, this special flavour seems in some ways to have acted against the transfer of information from the Genetics Institute to the rest of Germany.

Klaus Rajewsky

It may rather have generated a problem with respect to the transfer of people from the Institute to other institutions in the country. The science we did, and that must also be true for some of my colleagues, was mainly done with groups abroad and not with groups in Germany. It was only later, at the time when the *Zentrum für Molekulare Medizin* was founded, that this changed. I got a lot of complaints from the German students in the lab at the time, who said that they had no chance to enter the German scene because there were no contacts.

Ute Deichmann

I would like to say something from the perspective of a student of biology in the 1970s. I studied in Heidelberg and the Genetics Institute there was regarded to be very different from the "normal" biological institutes. It was regarded as American, and this was not a favourable attribute then, nor it is now. I imagine it was similar here in Cologne.

You mentioned that one of the special features of the Cologne Institute is the non-hierarchical, relaxed way of working, and I wonder whether this is really true. During my biology studies I experienced a quite relaxed atmosphere at Zoology and Botany Institutes. When I came to the Genetics Institute in Cologne (to begin my work in the history of science) and talked to students, I did not get the impression that the Institute's structure was particularly non-hierarchical.

Klaus Rajewsky

I very much appreciate what you are saying. The Institute started out as a place that was extremely non-hierarchical and that was a very special experience. It was like Benno said, that there were not many people in the Institute, but everybody who was there was basically working independently. I remember that students were circulating through all the labs and they knew everybody, including all the PIs [principal investigators], and they did their projects in discussion with many different people. This very special situation gradually changed as the Institute grew. With respect to what you say about the Institute being considered "American": although it may not have been ideal for some of our students, I enjoyed very much the special attitude of the Institute being orientated towards the world and not towards Germany.

Speaker from the audience

I would like to make a comment on the relationship at the time with the medical faculty because I remember, as I was Klaus' first Ph.D. student. At the time I was working on B- and T-cells collaboration in the immune-system. I remember when you presented data at a congress in front of clinicians and you told them that there were cells in our body that cooperated with each other; they did not really believe that this could be possible. They had never heard that cells cooperate with each other. That was so strange for me: to think of medical people who cannot imagine that there are cells in our body that cooperate with each other. I remember how you were trying to convince them of the consensus that, at least in the immunological system, you can see this cooperation, and it was just like feeling ignorance from the side of the medical faculty. I think it was very new in Cologne, to start this basic immunology, and, of course, it continued over decades from then.

Speaker from the audience

You said that it might have been a problem for Germans to get into the German system outside of Cologne. How difficult was it for Americans if they wanted to go back, because nowadays there are not that many American post-docs in Germany?

Klaus Rajewsky

That is an interesting point. In the early years we had quite a number of American post-docs, and that was also true for other institutions like the EMBL in Heidelberg and, a little later, the Basel Institute for Immunology. I think at that time it was not a problem for the Americans to go back. In fact, none of the guys who worked in my lab had difficulties going back; they all made their careers back in the US. Now, for reasons which I do not quite understand, the influx of Americans to Europe has dramatically dropped, which is an unfortunate development.

Walter Doerfler

In the 1970s and 1980s it was relatively easy to convince American post-docs to come to Cologne, but later on it was an exception that Americans came to Europe and Germany. The main reason was that they wanted to stay in the Harvard, Stanford, Rockefeller circle. And if you wanted to get out of it Cologne was, of course, a good choice, as was Pasteur. But it is not quite equivalent, at least in their imagination. The other aspect you mentioned was the collaboration. When

I came to Cologne in 1972 my main scientific ties were to groups in the United States because I had spent there many years. Of course I kept these contacts and we tried to branch out to the Dutch people working with similar ideas, and that worked very well. But we were also trying to keep the contact to German biologists, to the Giessen-group or the Heidelberg-group, which also worked out quite well.

Let me comment on what you said or what Benno had said before: I think I was only asked occasionally to fill biology chairs, so this was not unique to Benno Müller-Hill. I am not sure whether it was the political aspect of the Institute — I never considered myself to be one of these red leftists, but I was considered to be a little bit to the right (though my friends and family always thought I was quite far to the left). I do not think that this was so important, but it was the foreigners, the American influences that the Institute reflected to the German world.

Klaus Rajewsky

I just have to tell a little anecdote here because it corresponds directly to this point. Once, during my time in Cologne, I was in negotiations with Munich for a chair in the medical faculty. That was during the time when the peace movement was strong and my first wife was actually active in the peace movement. So the negotiations went on and it came to the final vote in the medical faculty in Munich. For the commission I was in first position for this professorship, but in the final faculty meeting the argument was raised that my wife was actually working in the peace movement, which led to a reversal of the listing of the candidates! This was in the middle of the 1970s or the 1980s.

Willy Stoffel

You stressed the perception and the attitude from the outside that the Institute was left wing. But that might be only partly true. Intellectually, here, inside Cologne, it was a liberal institute. But in general all members of the Institute were also conservative. I must say this because as a conservative myself I could not have felt at home if you were not of that attitude. In this attitude, this mixture of intellectual liberty and this rather conservative attitude in academic matters, I see also the success of the different groups.

Klaus Rajewsky

I am totally in agreement with you. I think the perception of the Institute from the outside was largely based on matters of style. However, the Institute was definitely a liberal place. For example, Benno's work on the "Murderous Science" was possible in the Institute. I think that it would hardly have been possible in any other academic institution in Germany at that time. And I think this liberal attitude was considered to be left wing.

24

Early Years of Transposon Research in Cologne

Heinz Saedler

More than 80% of my academic life was somehow associated with the Institute of Genetics. The first 12 years as a student, assistant and *Privatdozent*, and the last 25 years as *Honorarprofessor* while being a director at the Max Planck Institute for Plant Breeding Research, about seven kilometres away from the Institute of Genetics.

I now want to briefly describe some of the Ph.D. work I did with Peter Starlinger leading to the discovery of the IS-elements (insertion elements, see below) in bacteria in 1968. At that time the regulation of the galactose operon in *E. coli* was studied in his laboratory. Geneticists usually start programmes with a mutant hunt, in this case with the isolation of galactose-resistant mutants.

Human genetics had revealed that galactosemia (an inherited disorder characterised by the inability to utilise galactose) was partly due to the accumulation of galactose-1-phosphate, thus causing stasis of cellular growth. *E. coli* features a similar phenotype: bacteriostasis occurs if galactose-1-phosphate accumulates, thus providing a strong selection for galactose resistant mutants. Most of these mutants had incredible properties, i.e. they were not simply kinase deficient, as one would have expected, but the mutations were located in upstream genes of the operon and affected kinase expression; in other words, they were strongly polar. Strangely enough, they reverted spontaneously at a high rate, but did not respond to any known mutagen including mutagens, which caused frameshift mutations. Such unstable mutations could have been of three types: duplications, inversions or insertions.

Based on experiments by Gritt Kellenberger, who had shown that the density of bacteriophage lambda was mostly determined by the DNA content in the phage head, the nature of the strong polar mutations could be analysed. The density of lambda phages carrying the galactose operon with and without the polar mutations was compared in double labelling experiments. One phage population was labelled with tritium and the other with C-14, and the mixture, analysed in CsCl-gradients, clearly revealed a density difference from which the size of the DNA insertion could be calculated. Revertants of the mutation restored the density of the gal transducing lambda phage. Hence, the strong polar mutations were due to DNA-insertions of 500 to 2000 base pairs into the galactose operon of *E. coli.*

At the same time Jim Shapiro in Britain carried out similar independent experiments and also found that strong polar mutations indeed are due to the insertions of DNA. Years later Heinz Josef Hirsch, also a student in Peter Starlinger's laboratory, in DNA heteroduplex studies revealed that these mutations were DNA insertion elements of particular sizes, which hence were called IS-elements.

Where does this additional DNA in the galactose operon come from? Before the era of restriction enzymes we clearly demonstrated in DNA-DNA hybridisation experiments that these DNA-elements (IS1 and IS2) were integral parts of the *E. coli* chromosome and some of its plasmids.

Most people still thought that DNA insertion elements are an oddity and they did not pay much attention to this new class of mutations. The picture, however, changed when, in collaboration with the laboratory of Norman Davidson at Caltech in1972, we succeeded in showing that they were also components of R and F episomes. In F episomes they (IS2 and IS3) apparently provided DNA sequence homology for their integration into the chromosome to generate HFR strains. While on R-factors they demarked strategic positions to separate the R determinant from the transfer unit, as IS1 did, or embraced individual antibiotic resistance genes, like IS1 in the case of chloramphenicol-resistance or IS3 in the case of tetracyclin-resistance to name but a few, made these genes movable. That was quite interesting for clinicians who had serious problems with multi-drug resistant strains and the R-plasmid structure provided at least a way to understand how antibiotic resistant genes could be transmitted from one plasmid to another.

By bringing people from molecular biology together with people from the bacteriological laboratories of medical schools the field of IS-elements caught on and became very popular. Antibiotic resistance genes apparently moved via transposition, and that automatically coined the name transposons for these.

IS-elements, considered miniature transposons, because they can move on their own, also had some other striking properties. IS2, for example, transposed and, depending on its orientation of integration, could turn on or turn off adjacent genes, while IS1 was extremely deletogenic. Deletions were generated to either side of the integrated element at very high frequencies. These unusual properties were characteristic for these IS-elements.

Why is this worth mentioning? Barbara McClintock had described unstable mutations in corn often associated with genes controlling anthocyane biosynthesis, thus leading to kernel colour variegation. Further analysis led her to the concept of transposable "controlling elements" in *Zea mays* (corn), for which she received the Nobel Prize in 1983. They are called "controlling elements", because she was under the impression that they are really controlling the development of maize. Years later, we, and others, found out that this was not the case; integration into genes responding to developmental cues was accidental.

Since IS-elements from bacteria and "controlling elements" from corn shared so many properties, it was no surprise that Peter Starlinger from the Institute of Genetics and I, as soon as I was appointed director at the Max Planck Institute for Plant Breeding Research (MPIZ) here in Cologne, changed research topics from bacteria to the plant world. Both of us studied now transposable systems in plants. While Peter's group studied the Ac/Ds system of *Zea mays,* our group focussed on the En/Spm system of this plant. Both groups were successful in cloning and characterising these systems in the early 1980s. In a major project supported by the federal government, plant transposable elements were used as tools to isolate other plant genes. But this is another story, which mostly was done at MPIZ.

Peter Starlinger already outlined the history of the foundation of the Institute of Genetics, as well as the integral role that Professor Joseph Straub (dean of the science faculty, director of the Botanical Institute) played in this process. Upon his appointment in the early 1960s as director of the Max Planck Institute, the relationship of MPIZ and the University was initiated and further developed over the decades. Here are some of the hallmarks of that history. SFB 74, "Molecular Biology of the Cell" (1970–88) was a very important project. All three departments existing at the time at MPIZ participated. However, most important was the development of the Gene Centre at Cologne, because it provided substantial federal support for infrastructural measures and development of molecular biology at the Institute of Genetics and at MPIZ in the "red and green" sciences. The Max Delbrück Laboratory within the Max Planck Society was actually operative for 12 years, housing groups of complementary research interests to those at MPIZ and at the Institute of Genetics. Most of the group leaders later became professors either in the United Kingdom, United States, Singapore or here in Germany.

The interaction between the two red and green institutes is still alive, in contrast to the red-green coalition in politics, which was not so successful. As politics changes, science also moves on, and thus I can only hope that the new generation of directors at the Institute of Genetics and at the Max Planck Institute für Züchtungsforschung can revive this tradition and turn Cologne into as exciting a place as it was in the past.

VI. Molecular Biology and the German University Structure

25

Panel Discussion

Hermann Bujard, Walter Doerfler, Klaus Rajewsky, chair: Maria Leptin

---

Maria Leptin

This session is about the introduction of the American departmental system to universities in Germany — primarily here in Cologne — and whether Cologne served as an example. We have touched upon that many times already. And I think it has been very interesting throughout the discussion so far to see the power of examples. We have heard about our colleagues here wanting to introduce things like lab coats or no lab coats, English or no English, and authorships of junior members of the lab on papers. There seems to be a very strong power of what you've seen elsewhere and found good. Whatever the departmental system may be — we will see whether that can be defined — it is another such case. And so we have here two of our colleagues who used to be at this institute as professors and as heads of hierarchical or non-hierarchical structures; that, too, has been touched upon. And we also have Hermann Bujard who has been responsible for founding probably the only other, perhaps even the only, German University Institute in which a truly non-hierarchical and departmental structure is embodied in the concept of the Institute. We will first have Walter Doerfler give us some ideas on this topic, Hermann will then talk and Klaus Rajewsky will make his comments at the end. We will have just brief statements. Then we will have a discussion in which we will allow everyone to reply to these points, and then we will take input from the audience.

Walter Doerfler

In reflecting on the acceptance of molecular biology in the German university structure, I am strongly influenced by experiences during my years of training in Germany and in the United States. There have been five different phases, fortunately all good experiences. The *Anatomische Institut* in Munich, where I did a thesis with Titus von Lanz, was my first exposure to research, although to a limited extent. Professor von Lanz was a strong personality and I owe him gratitude for the many hours of formative discussions and the good advice — among many others, "*Gut Ding will Weile*".

My next impression was in Munich with Wolfram Zillig. He was at that time an independent group leader at the Max Planck Institute for Biochemistry under the directorship of Professor Butenandt. There were several truly independent groups in the Institute. The atmosphere was very liberal, although Adolf Butenandt came across as a strong but very encouraging teacher. Zillig's group was totally non-hierarchical and appeared American in spirit.

The most important role model for me, of course, was the department of Biochemistry of Stanford University, with eight outstanding faculty members and not that much else in infrastructure. The department operated in an admirable fashion mainly on grants and training funds in the good old days of the NIH (National Institute of Health). There was no question that all the group leaders were totally independent of Arthur Kornberg although the atmosphere and style of the department, to which all faculty members contributed, had originally been created by him. The Stanford Biochemistry Department to this day appears to me the ultimate example of how to run a university institution. Karl Muench, one of the post-docs with Paul Berg, predicted at that time, "from Stanford you can only go downhill". A strong statement, but Karl was correct in a way.

Next, I was most fortunate to receive an attractive offer to join the virology group at Rockefeller University in New York City. Dave Hogness, my advisor at Stanford, had primed me with: "the moment you walk into Rockefeller, you make it absolutely clear that you are on your own". I was again fortunate that the virology group headed by Igor Tamm and Purnell Choppin was not only not hierarchical, but very open, liberal, and most helpful in many ways in supporting my research. Igor Tamm was a gracious, gentlemanly person, and chair of the department. My declared intent to work with adenovirus in an RNA virus group initially raised some eyebrows but was perfectly accepted and everyone in the group was kind and supportive. The atmosphere at Rockefeller was very much group-specific, and I soon learned that I had made a good choice in joining the Tamm-Choppin laboratory. The structure of this research university — with a faculty to student ratio of 4:1 — was geared to maximise the effectiveness of good science. The administration at all levels was meant to work for the scientists at the Institute. The ordering department, directed by Tony Campo, and the entire infrastructure at Rockefeller was a legend in know-how. If you knew what research you wanted to do, Rockefeller was a scientist's paradise.

After having accepted an offer from the Institute of Genetics in Cologne, I spent almost a year as visiting professor in Lennart Philipson's group at the Wallenberg Laboratory of Uppsala University. I learned a lot from Lennart, of course, through our collaborations, which had already started in New York, but I mainly learned how to organise a university department with many young students in a European environment.

Re-entering the German university system that I had known only as a student, after many years abroad, led to the unavoidable culture shock. Although the Institute of Genetics in Cologne had maintained its US-style atmosphere and structure even after Max Delbrück's departure, the university lacked many features that made Stanford and Rockefeller such unique places. The only way to survive under these circumstances was to organise my own group according to what I had learned the preceding 12 years, and ignore the German university system as much as possible. Of course, there had to be compromises; it would have been senseless to confront the clumsy system head-on. Again, one of Arthur Kornberg's adages was helpful: "avoid trouble, don't fix it".

To this day, I am firmly convinced that in Germany's system of higher education we would have done ourselves a big favour if, instead of importing Coca-Çola and McDonald's, we had introduced the American departmental system to our universities already in the 1950s or 1960s. Of course, I quickly learned that this was a totally naïve perspective, and practically no one at the University of Cologne would have been willing to pursue that path. So, the Institute of Genetics stood solitarily alone in such endeavours. Nevertheless, it would be unfair not to acknowledge that during many years we had excellent support from the administration of the University of Cologne. Above all, without the continued support by the *Deutsche Forschungsgemeinschaft,* I could never have come to or stayed at the Cologne Institute of Genetics. The best thing I encountered at the University of Cologne was the unabated interest and sustained motivation of the young students and independent group leaders whom we succeeded in interesting in molecular genetics and in basic research.

As a result of German labour laws, which have not been adapted in decades to the needs of universities, one of the major assets we are still missing in the system is a tenure track structure for junior faculty. Over the years, we, unfortunately, were unable to attract many a promising young investigator to the Cologne Institute for exactly that reason. There have been several dilettante attempts at *Hochschulreform* handed down by politicians. My advice: introduce a tenure track system for junior faculty and do everything possible to attract the best possible scientists to the faculties, and everything else will follow. The current downgrading of faculty salaries will prove to be yet another demonstration of ignorance of encyclopaedic proportions.

Hermann Bujard

Let me first comment on the remark that the Institut für Genetik in Köln was isolated or felt isolated in a unique way. This was probably true during the initial phase. Thereafter, it was certainly not a unique situation; rather it was a general phenomenon experienced by those of us who returned from the US and brought along a new discipline, molecular genetics or molecular biology. We entered the faculties of biology where the classical disciplines like zoology, botany and microbiology were fostered — usually at a level which was not very impressive. Indeed, in such faculties, we were isolated as we did not only ask different questions in our research and teaching, but we also had a different working style and we dealt with our students differently (no title, open office). On the other hand, within Germany, I never felt isolated after my return to Heidelberg. There were the Institut für Genetik in Köln; the MPI für Molekulare Genetik in Berlin, with Thomas Trautner and others; the MPI für Biochemie in München, e.g. with Wolfram Zillig and Hans-Peter Hofschneider; and a number of other groups throughout Germany. We all knew each other pretty well and had interactions. For example, in Heidelberg, we had Jeffrey Miller teaching genetics while he was at Köln and we got Fritz Melchers from Basel to teach immunology at our place. Thus, while we were isolated in a way within our faculties, there were fruitful scientific and personal relationships among this generation of molecular biologists. They were rewarding and compensated for our troubles in our faculties, which were often driven by the insecurities and inferiority complexes of our colleagues in the classical disciplines. For me, those times were exciting for very special reasons; after all, within 10 to 15 years, we had a good impact on the teaching and research in our previously pretty old-fashioned faculties!

Now, before discussing the department system, let me make some additional remarks. During my years spent in the United States, first as a post-doc and thereafter as an assistant professor, I was most impressed by three specifics which structured our life in the academic world:

- Its policy of giving grant support to individual researchers is, in my view, the most important and most successful way of fostering good research. It is most disturbing that the Deutsche Forschungsgemeinschaft has dramatically reduced this program and instead pours money into doubtful "networks", "research schools", and "SFBs", all of which are prone to support mediocrity and mainstream ideas. Needless to say, the EC granting system is even worse.

- In the US, more than 85% of basic (and also much of applied) research is carried out at universities, where the next generation of researchers is exposed early on to first-rate scholars and their science. By contrast, we afford increasingly large — and rich — research organisations outside of universities, as exemplified by the Max-Planck Institute, the Helmholtz Institute, and the Leibniz-Gesellschaft, to name just some major organisations. For obvious reasons, these organisations neither contribute sufficiently to the education of the next generation of scientists nor do their members in my view excel in original research accomplishments that would justify the privileges they enjoy.
- US universities are structured in departments which, in my view, provide the best prerequisites for research and teaching as well as transparent academic careers for the young and talented.

It is the department system that makes universities in the US superior to the standard European university. Now, what are the specifics of a department?

Firstly, its faculty consists of several "equals", i.e. full professors. They are joined by several independent associate professors (generally with tenure) and a proper number of assistant professors (not tenured young scientists). All group leaders are fully independent with respect to their research interest, but at the same time are also fully responsible for their own success. In a well-organised department, one shares common resources and responsibilities of the department, e.g. in teaching. These features are the key to quality control among peers: since you like to share the common resources only with the best — and not the mediocre — colleagues, you seek respective faculty members who are also prepared to share responsibilities. Such departments generate a *Mehrwert* (added value), as their faculty strives to hire the best possible candidate for an open position. This striving for the best colleague is usually even supported by weaker members of the faculty, as they would also profit from an increased reputation of the department and sometimes even from an increase of incoming grant money. The result is a corporate identity that penetrates faculty and students, and that constitutes the most important "first level" of quality control. It is this spirit which is generally absent in *Fakultäten* of German universities, where the most influential members (full professors or *Lehrstuhlinhaber*) are primarily interested in defending their own *Lehrstuhl/Institut*. After all, *Besitzstandswahrung* has reached the status of a human right in our society!

Secondly, the department structure as sketched above provides the most appropriate framework for giving young, talented scientists a chance to develop independently. It allows the establishment of a transparent tenure track procedure for assistant professors who are independent young colleagues within the department, and not senior research associates within the group of a full professor. Accordingly, promotions (e.g. to a tenured position) will be the result of a decision by the department's faculty, which, as defined above, is of course interested in keeping only the best.

Most importantly, a department profits greatly from a certain turnover of young assistant professors, who bring along new ideas and technologies, particularly when the department's policy encourages young group leaders to carry out "offensive" research, i.e. research outside the mainstream. Such a policy however, requires us to defend and support such young colleagues should they fail, given that our present-day shallow evaluation system rates impact points more highly than originality.

Finally, a department structure can integrate colleagues who may fail in organizing a big *Lehrstuhl/Institut*, but who are nonetheless real assets in a departmentally organised institute because of their outstanding scientific talent.

So, why does the department structure not "catch on" in European universities? In Germany, it would have to replace a formerly successful system, from the days when universities were about one tenth the size of today and where the *Lehrstuhlprinzip* functioned pretty impressively. In the early years of our universities' expansion, a *Lehrstuhl* or institute would also grow, but maintain its overall structure: the head of the institute would still be the full professor (*Lehrstuhlinnhaber*), but now with even more power. Thus, in Germany and other European countries, large hierarchical structures were generated. By contrast, in the US, the expansion of universities resulted generally in the departmentalisation of institutes. One has to be aware, however, that some of these classical European institutes, depending on the professor, were still remarkably successful. Therefore, they are often used as paradigms for an alternative to the US-derived department, sometimes even with the nonsensical argument of different mentalities on both sides of the Atlantic. The true dilemma preventing necessary changes, however, is that present-day *Lehrstuhlinhaber* (full professors) would have to give up formal power and would have to concern themselves again directly with research when entering a departmental structure — and again we are confronted, not only in Germany, with the holy cow of *Besitzstandswahrung*.

So what can be done? We have to work on insight into the problem at the political level, with the goal of providing resources that can be used as incentives for colleagues who are willing to join a true department structure. It has to be made clear to decision makers at the political level that such a reorganisation of research and teaching units will lead to a more efficient utilisation of taxpayers' money and will create a true *Mehrwert* in education and research.

Klaus Rajewsky

I think what Hermann Bujard said about the individual grants is an absolutely essential thing. Looking back from Harvard on the scene here in Europe, I find it shocking that it seems you have accommodate yourself more and more in scientific "networks" or the like in order to get your grants. In America this is really different. If you have a good idea you get your grant and do what you want to do. There is a second thing I want to add to what Hermann said. Of course, it is easy for me to say this after having enjoyed the situation for so long, but I think a main element for a true department system is to take away a lot of the privileges which the established professors in this country are enjoying. During their career the professors get more and more positions and are granted money, and they don't want to share that with the rest of the department. That is a fundamental mistake.

I think most of the available money and positions should be put into a pool for the department and then given, on a non-permanent basis, to the people who really deserve them by scientific merit. By the way, that is also the only way that the European system can get rid of obligatory retirement because, of course, nowadays young people ask for how long professors should have all those privileges, as they do also want to have them for themselves at some point. In America either you get your grant or you don't and that's what counts. I think that is an essential feature of a truly functional departmental system.

The Institute of Genetics has done a pretty good job in the framework of the existing structures, in the sense that the Institute has managed to preserve the idea that the main thing that counts is the scientific argument. In the end we have always made decisions on the basis of science and not on the basis of administrative considerations. But, of course, we were limited in our efforts by the general structure in which we worked. Thus, we really exploited the instrument of the five-year groups as a means to attract young group leaders to the Institute (my department had 13 such research groups during its existence). But this instrument was severely hampered by the five-year time limit, which made it impossible to establish a tenure track system. We didn't manage, like you did in Heidelberg to some extent, to resolve this issue.

Maria Leptin

We have heard a few positions and statements here and I think we will just go through the panel and allow each of you to reply to what has been said. Then we will take urgent discussion points, and then we will have the general discussion.

Walter Doerfler

For a country like Germany, the paramount goal must be the support of innovative, investigator-initiated scientific research. There has been an increasing tendency to almost force scientists into support structures of collaborative projects of larger and larger groups — *Verbundforschung* — as the juggernaut expression stands for apply-side oriented work, preferably of medical relevance. While there can be no doubt that molecular genetics has had a tremendous impact on medicine, prescribed goals—cancer research, HIV work and others — will profit most likely not from strictly oriented projects, but rather from basic research that strives to solve a fundamental biological problem. Politicians — in Berlin or, even worse, in Brussels — plainly lack the required expertise to advise or persuade scientists towards the type of research or the partners in collaboration they know best to select on their own.

Brussels is full of bureaucrats administering science grants. I once served on a committee of the EU, travelling around Europe and evaluating projects in medical genetics. The young investigators in London or Paris more or less ignored us and would hardly talk to us as representatives of a bureaucratic system they abhorred. When I later related that experience to one of the bureaucrats in Brussels, he retorted: "Well, young man, relax. I have been involved in this administrative work for thirty-five years, and I know best how to manipulate it." He obviously was far removed from reality. This anecdotal excerpt from my travels for the EU — I since rejected with thanks similar offers from Brussels or Berlin — appears very typical of how the system functions. There may be hope for the future.

In Germany, we have to be careful that the DFG will not be overburdened by requests from politicians to receive increased support on the premise of directed research on this or that "worthy goal". It appears that the *Normalverfahren,* the essential means of support for investigator-initiated projects, finds itself stripped of funding in favour of exactly such "worthy goals". The DFG is severely underfunded in general. The utopian project inaugurated by politicians to create *Elite Universitäten* overnight by spending a few billion euros again exposes the poor quality of our political class. Of course, the money may be helpful to the lucky few who qualify in the questionable selection process. On the down side, this morose political initiative should not prevent us from demanding, in a much more intellectually aggressive way, expanded and long-lasting funding for the universities in Germany and for the granting institutions like the DFG.

Hermann Bujard

I would like to come back to the point of why a departmental system is so important. Walter Doerfler mentioned Arthur Kornberg, who said that he was at least as proud of having formed his department as of many of his publications. This is an important statement, as departments need chairmen. In Germany, we are missing the chairman culture you have in the United States. A chairman should have *Gestaltungsmöglichkeit,* and of course his goal should be to hire the best scholars, who sometimes might even be better than himself. This needs a certain type of individual. In the US, such individuals can, when successful, enter attractive careers: successful chairmanship is often a qualification for becoming a university president, a presidential advisor or other important roles at the interface between science and politics. For example, David Baltimore, for years the chairman of the Whitehead Institute at MIT, became president of Caltech; Shirley Tilghman, president of Princeton University, was formerly a successful chairwoman of the biology department; or look at the careers of Bruce Alberts, Harold Varmus, and Mike Bishop, to mention just a few. In our universities, we do not have the culture for such impressive careers as our capable colleagues. We very rarely have successful researchers in natural sciences who would agree to become *Rektor* of a university, and it is a disaster that we usually leave this job to law professors and historians. Thus, as long as we do not develop a department system, we will not foster chairpersons who become visible by doing an excellent job and who enjoy engaging themselves in science at a different level with the promise of a successful and internationally recognised research career. These individuals are urgently needed for taking over responsibilities in the European Community, the DFG and various research organisations.

Klaus Rajewsky

I just want to add that you cannot have these people if you don't have the legal structure which gives them the necessary power. We have to declare that the constitution would have to be changed to get that to work. You got it to work through a miracle of convincing your colleagues that they should all act together to do this. Legally speaking they could all have blocked it from the beginning and it would not have worked.

Hermann Bujard

For the ZMBH in Heidelberg, we have exploited two fortunate situations. Firstly, during a decisive period, we had a *Rektor*, a physicist who allowed us to establish the institute outside of the *Fakultät für Biologie* and to report directly to the *Rektor* as a *Zentrale Forschungseinrichtung*. Secondly, Heinz Schaller and myself refrained from having formal chairs and we subsequently did not offer chairs to colleagues who liked to join the ZMBH, in accordance with our institute's bylaws, which we established without breaching the Universitätsgesetz of the State. That is how we got it rolling, but I agree with you that it is extremely difficult as an individual professor to bring such change about at a German university.

Maria Leptin

I think that there is a certain consensus here. We all seem to know what we want. We want money to be distributed differently from the way it is done. There should be more individual grants and it should not be under the control of politicians. We also seem to agree that structures need to be changed. Of course, some good things are happening, such as the European Research Council, which will be created and will be under the leadership of people who have very much the same priorities. So we should not be too pessimistic. But perhaps my colleagues here would like to say something about how they feel about this in terms of practical considerations. What do we do? We all complain, but why is nothing changing? Or is it a misconception that we all agree? Is it just that the people who are here happen to agree? Maybe we should have some proposals and some ideas before we continue to debate.

Walter Doerfler

It is always difficult to offer realistic proposals for improvements in the German university system, but as I mentioned before, there would have to be many more unified and convincing efforts by the scientific community to educate our politicians. That is not easy to do as I can demonstrate to you by one example. In the early 1990s, all the directors of the Institute of Genetics and the Max Planck Institute for Plant Breeding Research tried to convince Herr Catenhusen, then chair of the Science Committee of the Bundestag, that the gene technology law would be unnecessary and would prove harmful. Herr Catenhusen in a memorable session this afternoon divined to us: "*Was immer Sie in Vertretung von Eigeninteressen sagen, wir im Bundestag treffen die Entscheidung*". How "lucky" can a country be to be run by such politicians?

In 1992 I spent a lot of time talking to politicians then still in Bonn, trying to convince them that they should mitigate the gene technology law. To support that, I asked all the German EMBO members to write letters to the then German chancellor Kohl. I only got three answers: from Lennart Philipson, director of the EMBO laboratory in Heidelberg, Eberhard Wecker in Würzburg, and Otto Westphal in Freiburg, who wrote strong letters. Of course, you can question this naïve approach of writing letters. But three letters from nearly one hundred members will not make an impact. Later on, I decided to concentrate my energies on science instead of on politicians. Perhaps, there will be a day when young scientists will try to go out and talk to their local representatives and to seek contact with politicians in Berlin. I am not at all optimistic that this will yield results in the short run.

Hermann Bujard

The real question is how such a fundamental change like departmentalisation of a university with all its positive consequences can be brought about. It cannot and should not be done by force, as professors have valid contracts and it would cause an endless struggle to pressure them by legal or administrative measures. Instead, one would have to generate incentives — a strategy I have unsuccessfully tried to convince the Ministry of Research in Baden-Württemberg for years. If proper awards are in sight for professors who would be willing to join a well set-up department, I am convinced that numerous colleagues would join. After a short while, they would realise the advantages of a well-functioning department and it can be anticipated that, given the proper incentives, the new structure would become contagious! No law at the level of our parliaments would be required for such a reform. All that such a reform needs is the will on the side of the professors and some money — actually ridiculously little compared to the gain — from the government.

Klaus Rajewsky

I am thinking of two ways in which a good Institute could be organised. One is the example of the MRC (Medical Research Council) in Cambridge that is the Laboratory of Molecular Biology, which works in a different way than most other institutes. Basically the groups in that institute get money for current research and are granted technical services, but they don't get positions other than technical ones. So they all live on good post-docs who apply to the labs, get their fellowships and then work with the principal investigators. For me it seems like a revelation, how well that works. It guarantees a good quality of research, because if it were not good, they would not get post-docs. So I wonder whether the DFG couldn't set up a system like that, ensuring that research groups have sufficient money for research expenses, but limited funds for scientific personnel, so that they would have to attract post-docs bringing their own fellowships.

The second way is the idea of giving people sufficient money in terms of individual grants like in the United States. This is an extremely healthy way to do because the institutions have to structure their operations on that basis.

Maria Leptin

There is one thing, of course, that we have not touched upon that you mentioned because you chose the Laboratory of Molecular Biology in Cambridge as an example. It is a research institute and we are talking here about a university institute. We will have a discussion on teaching this afternoon. Of course, many of the posts that are given to us and that this Institute has very clearly used often for independent scientists are the *Assistenten* positions. They are partly given to university professors to help them with teaching, a function that is needed and that has to be thought about in this context. So we haven't discussed that point at all.

I think it is time to start an open discussion and encourage comments from the audience. I would particularly like to also ask the younger people here to let everyone know their views. We have heard a lot from us here on the panel and from very senior people but there are fortunately some junior people here. So please do feel free to give us your comments on the questions that we have been discussing. It would be nice to hear from you.

Joseph Lengeler

I think it is very important to distinguish between comparably famous and large German institutes like the present one or the one in Heidelberg and the average small German institutes. Furthermore, we have to consider that the official functions of the professors in top-class American universities, which we are supposed to copy, and those of the average German professor are quite different; the latter having a much heavier teaching load and more administrative obligations. When I started in 1984 in Osnabrück, we had no "department" (*Abteilung*) for botany, zoology, or whatever, but a very modern *Fachbereich Biologie/Chemie*. It comprised 10 independent research groups representing the various levels of complexity, such as biophysics, genetics and ecology. We were 11 assistant and full professors, teaching one curriculum together, *Diplombiologie*, with about

90 new students per year and 20-odd prospective teachers on a different curriculum. One full professor and three post-docs had to teach genetics, which meant, besides lecturing, the training of about 40 selected (interested) students in practical laboratory courses. Although the university was only founded in 1975 we established from 1985 to this day two very successful *Sonderforschungsbereiche* at the Institute. Now, in 2006, there are 220 to 260 new students per year taught by 12 professors, who are additionally forced to introduce two bachelor and master curricula. Inevitably, the teaching load increased. At present, one full professor and three post-docs have to teach genetics for about 80 non-selected, mostly non-interested students in practical laboratory courses 12 months per year. Finally, small universities often lack medical or engineering departments or the neighbouring MPIs which would help them to establish well funded collaborations. In Göttingen and Munich, for example, a major percentage of the Ph.D. students in biology are funded by the medical departments or by MPIs. This "average" situation, together with the lack of adequate funding mentioned before, represents reality for the majority of German biologists. When comparing the scientific results of such groups with the results of the better funded groups, the former very often do not deserve the criticism it receives — which we also heard during this discussion. Just the opposite!

Ariane Toussaint

I know the Belgian and the French university systems and everything that I have heard so far is true for both of them. I think that we are really dealing with a European problem here. I don't know about the English system but it concerns at least these three countries. Regarding the difficulties in the contact with politicians I know that when you study in France you go to the *Grandes Écoles*, and people who go into politics do not go to the same schools to which scientists go. They go to the *École normale* or *Polytechnique*. So, I wonder whether a similar system exists in Germany, and whether that might not be a reason why it is so difficult to communicate with each other as these people have a completely different background. In Belgium we don't have this problem but still communication is very difficult.

Remembering the old days, I also want to mention the fact that Peter Starlinger came to Brussels in 1968. At that time we were fighting with the university authorities, trying to get rid of them and to create a department of molecular biology at the University of Brussels. There was a department of zoology and a department of botany. They existed until last year; they are now the department of biology. So there is a department of biology and molecular biology but we still haven't solved the problem of competition between biology and molecular biology. This is really a European problem, just like the finance problem.

Ute Deichmann

I want to go back to the beginning of the Institute of Genetics here in Cologne. My question is: why was it so important to Delbrück to implement the departmental system in Germany? What exactly did he mean by trying that, and was he successful? The background for my question is that Professor Bujard said that the old German system was so exceptionally successful before the war. I know that there was some great success, such as the work of Emil Fischer in chemistry, but I also see two major flaws already at the beginning of the twentieth century. First, there were the difficulties in establishing new scientific fields at universities. So there were almost no biochemical university institutes in Germany, despite the fact that the best biochemists worked in Germany. They sometimes worked in hospital labs, like Leonor Michaelis (he is best known for having derived, with Maud Menten, the affinity constant of the enzyme substrate bond) who therefore left Germany in the 1920s.

Second there was the strong hierarchy and the powers of a professor. Hierarchies are necessary but in many cases they turned out to be harmful. An example is the case of Emil Abderhalden, professor at the University of Halle, whose reputation was based on research starting around 1910, which was soon shown not to be reproducible. But he was able to continue with that until he died in 1950, and he was even made president of a major German Academy of Science. His post-docs and graduate students were afraid to say that Abderhalden's method would not work. I guess that this strong hierarchy is one thing that Max Delbrück wanted to change. My question is: did he succeed?

Hermann Bujard

Back in the days when the "old" German system was successful, it did not differ dramatically from the situation in the US. Full professors were powerful, but their institutes were relatively small. The diverging developments started in most fields in the 1950s and 1960s, when universities expanded. Germany was, after the war, unfortunately in a very restorative phase. I am sure that not too many people were taking the time to think about more appropriate structures, particularly if they would lead to less hierarchical organisations. So, I guess one aim of Delbrück's initiative was to implement the liberal department structure, which worked so well in the US.

To establish a new discipline has always necessitated convincing the existing disciplines to more narrowly define themselves. In Germany, chemistry in academia was a strong discipline, but it was heavily influenced by the German chemical industry, for which biochemistry was not a priority. So, I believe that chemistry/biochemistry is a special issue in this country, and is still a touchy topic in many chemical faculties. On the other hand, if one follows up the development of physics, I do not believe that we would find the same situation.

Coming back to Abderhalden: if one were to analyse today's fraudulent cases in research in Germany, an overwhelming number would be found in large groups/institutes with excessive hierarchical structures, which are of course helpful when it comes to covering things up. So, we have another argument for a liberal department structure with a flat hierarchy.

Klaus Rajewsky

I wanted to ask Ute Deichmann whether the lack of flexibility in the old system could be related to the fact that in the old days professors had collaborators for, say, 15 years rather than for three to five years as is presently the case with post-docs. I think the flux of people in the institutions, with the young people becoming independent relatively early, is an essential feature of the departmental system. That may have been very different in the old structures.

Ute Deichmann

This is a very good point. For example, look at the *Habilitation*, and the fact that it took so long to become independent, and if you contradicted an authority you ran the risk of not receiving a professorship. To mention an example from the United States already in the 1930s: protein chemist Max Bergmann at the Rockefeller Institute in 1937 published his erroneous results that proteins had a periodical structure. Shortly after, in 1939, his young co-worker William Stein published that he could not reproduce Bergmann's results. He dared to do that and he did not have problems advancing in his career. This would have been impossible in Germany.

Walter Doerfler

Comparing the old German to the American system, many colleagues and friends in the United States told me that Johns Hopkins Medical School was set up according to the example of the Charité and the medical faculty in Vienna. Or the Rockefeller Institute for Medical Research, as it was called between 1901 and 1950, was a hybrid between Cambridge and the old German system. It still reflects many originally good ideas that have been weakened by bureaucracies, but could and should be revitalised in model institutions in Europe.

Speaker from the audience

This is probably more a comment than a question. I understand that funding is probably decreasing rather than being increased. So my question is: why not involve the pharma industry in funding the professors at universities? Probably this is difficult for professors who are answering more basic and fundamental questions in biology, but it would be worth trying it out.

Hermann Bujard

May I comment on this? Firstly, the pharmaceutical industry in Germany is no longer what it used to be: a worldwide leading prosperous branch (see e.g. the fate of Hoechst, Boehringer Mannheim, Knoll AG). Accordingly, there is much less potential support to expect. More importantly, however, in most (if not all) of these companies where the insight has faded, support of basic research is essential for prospering applied science. This is the consequence of a changed culture within these enterprises, where purely business-oriented individuals with little understanding for research have increasing influence. In the long run, this is a dangerous development.

At the ZMBH, we enjoyed support from BASF, Ludwigshafen, for 10 years, with practically no strings attached. Despite the success of the ZMBH, the BASF was not willing to continue its support unless they would have some say in what type of research would be carried out at our institute, a condition we declined. In my view, a university institute should never accept support from industry that exceeds around 5% of the budget, unless there is a long-term agreement in place to secure the institute's independence.

Jonathan Howard

I would like to comment on what Klaus Rajewsky said, which has to do with the level of resources which are taken for granted in the German professorial system. Strangely enough, it is one of the best kept secrets in Europe. That is because of the language barrier; the efforts that have to be made to understand what a German professor says. It is an interesting situation. I suspect that the German, the Austrian and the Swiss systems are rather similar and they assume that professors will be given something which is called a department. There is an underlying confusion about this *Abteilung*. The *Abteilung* is often translated as department, but a department is completely different, as it is a consortium of professors working together in organisation and teaching. So if everybody else knew what the resources allocated to German professors were, then they would all wish to take a job in Germany. But this turns out not to be the case.

I know the situation in Britain. When I first came here and only after I began to understand a little bit of what was going on, I started to understand what some of the structural problems were. I began to make some numerical comparisons between the resources in the faculty of biology here and the faculty of biology in Cambridge, where I came from. One very obvious number is that the average C4 professor in the Institute of Genetics has got three *Assistenten-Stellen* associated with his chair. In the British equivalent department of Genetics in Cambridge there are about 20 fully independent, independently funded and academically wholly autonomous groups. Now if I was to take away those three *Assistenten-Stellen* and turn them into independent positions, whether tenured or not yet tenured, I would create a situation very much like that of British universities.

I suspect that this is also something like what Klaus is experiencing in Harvard or colleagues have experienced elsewhere in the American system or the Anglo-Saxon system. I wanted to stress this because nobody has picked up this point and Klaus is very brave to say it. It is a fundamental structural issue because those positions which are held by the senior professors in the German system are the very positions that the people whom we nurture in our *Abteilungen* wish to have as independent scientists. But they don't get them, and there are complicated reasons for that. In reality the competition for independent academic positions in the university system should, in my view, be a competition for just those posts, but it is not. It is as simple as that.

Hermann Bujard

In principle, I agree with Jonathan, even though I see problems with some of his extrapolations. If you have a full professor position e.g. in

Baden-Württemberg, you have a personal teaching load of nine hours of formal teaching per week at the graduate and undergraduate levels. For me, there is no way to fulfil these duties and at the same time carry out research at a respectable level unless you have some personnel. Now, coworkers sponsored by research grants are not supposed to jump in, even though we regularly (mis) use them. This explains why, e.g. in the German system as it stands, professors need some personnel who, when it comes to *Assistenten*, in turn have their own teaching load that comes with the position. Clearly, the way out has to be the conversion of several *Lehrstühle* into a department where all positions are first pooled and then redistributed. Only in this way can one generate positions and packages for assistant professors and an infrastructure for technologies and teaching, accessible for every research group leader. What remains unsolved, even under these conditions, is the excessive teaching load and the highly unfavourable ratio of faculty to students.

At the ZMBH, which was formally created by joining six full professorships (*Lehrstühle*), we generated — in addition to the full professorships — three positions for associate professors and four positions for assistant professors. Around 50% of all resources (personnel, running costs) support our infrastructure, which is particularly valuable for the assistant professors (junior group leaders) and independent project leaders.

To sum up, I feel that it is dangerous to give politicians the impression that professorships are unduly oversponsored — don't forget, at least in Germany, universities are the *Armenhäuser* in our inefficient research landscape, in which we afford organisations like the Max-Planck, Helmholtz, and Leibniz-Gesellschaft, all of which are much better financed. Instead, we have to defend the little we have, but we have to make better use of it by restructuring our organisation. Not to carry out these overdue reforms is a historic failure of European universities — except again institutions in Great Britain.

Maria Leptin

I think this is a point we can continue with in this afternoon's discussion on teaching. Willy Stoffel has been waiting for a very long time to say something and then there is another young person who wants to say something. And if we then have time we can come back to this again.

Willy Stoffel

I think we have to focus on our own University because we cannot change politicians in short terms.We have a new *Rektor*; he is a physicist and very much in favour of all institutions, which contribute to the reputation of this University. So my suggestion is to bring together all scientifically active professors rather than changing constitutions, change the thinking in the University, and make clear that the reputation of our University is depending on these people. They have the particular weight to change the situation. We still have, in my opinion the best post docs, whom we sent to the US and they belong there also to the best.

So the problems are not our young scientists, but the distribution and control of money. What you said about the SFB is only partly true, because I think that mediocre and mainstream researchers are hidden mostly in SFBs. Individual research is much easier to evaluate than mainstream and mediocre research in big units. I have evaluated many SFBs and experienced that it is hard to find out the mediocre research which is hidden under the good research. There has to be some mechanism to be implemented in the DFG funding system to move some money out of the SFBs into the individual support mechanisms. More money should flow into the Universities.

Maria Leptin

Not much needs to be said, as we all have high hopes for the *Rektor*, and hope he lives up to them. But you are right that this is the time when, at least in Cologne, we can make progress in the right direction, and it is clear that the *Rektor* wants that. So we are running through open doors about redistribution, rewards and actually taking resources away. It is terribly painful because nobody wants to be the first to have resources taken away, but there is now a political will to narrow the power of individuals and distribute the money in a fairer and wider way. So we must hope that this gets us somewhere.

Willy Stoffel

What I mean is that supporting the *Rektor* also means to restrict bureaucracy. There is a permanent fight between the chancellor and its administration and biomedical research in Cologne. If we cut bureaucracy down to one third then we will be better off.

Speaker from the audience

I don't understand why biologists are so reluctant to consider the position of chancellor. As someone said before: that three will not have an effect, but three-hundred will make an impact. I believe that if a powerful chancellor was a biologist, a physicist or a chemist he would only have a few administrative tasks.

Maria Leptin

First of all we should explain to non-Germans that in Germany the names are the other way round. The chancellor is actually the administrative head who has a fairly powerful position, as he holds the money. The academic head of the university who determines policy is the *Rektor.* The chancellor is not elected, but is a career administrator. The *Rektor* is actually at the moment a physicist, which is very good for us, as Professor Stoffel has just said. It is just as good as having a biologist there because he does represent our ideals and ideas. And why the biologists are so reluctant is because basically you cannot run your research at the same time. And so most of us have not considered that position for that reason alone, as even being dean is a tremendous burden. That is the sad explanation.

In the American system somebody of that level would either be dean or just the chairman of a department, as those jobs are rewarded not only financially. That is not necessarily the reason why anybody would take that job but more in terms of glamour and recognition for what you do. Here in Germany a dean does not do that, as they are here for four years and then they are gone. If they screw up, well, then tough luck, as their successor can sort it out. If they do something really good then it may be recognised by a few people, but you have neither the responsibility nor the recognition and rewards that you get in other systems. So in my view there is too much democracy and too much administration in the system. I don't know whether this answers your question and I don't know whether my colleagues agree.

Gerrit Praefcke

I do not think that the government will have that much money in the near future and maybe even not in the more distant future. But there is a difference if you compare continental Europe to Britain and the United States. And this difference is that there are people who give huge sums of money: private people who donate it when they die. Rockefeller University does not only have its name because Rockefeller was a famous person, but also because he gave money. Or Vanderbilt or the Wellcome Trust. There will be huge amounts of money which will be inherited within the next 10 or 20 years. Maybe we should work on politicians to change the laws to make it more attractive and more rewarding for people to donate money to universities. Maybe we will have an "Aldi University" in five or 10 years. Just think of how much money these people have and how much money they could give to science.

Jonathan Howard

I want to follow what Gerrit Praefcke said. There is a huge deficiency in the German system which one notices more and more as time goes by, and that is that there is essentially no charity component in research funding. I am not talking about just putting your name on an annual fellowship, but much larger charity research funds, like Cancer Research UK, the Juvenile Diabetes Foundation, the March of Dimes, and so on. The list in other countries, and not just the US and UK, is very long. These are really major contributors to the delivery of science. I cannot think of anything of remotely comparable scale in Germany. Why not?

Frank Sprenger

I have the impression that it is very difficult these days to do high-risk science. Yet high-risk science might be important to produce good science. So I think one should also work on getting a system established where especially young people could do risky projects which work or might even not work. The problem is that if they don't work then you are on the street, as there is no safety net for people who are in the system for some time. We have nothing like in the United States. The evaluation of full professors is mainly done on research; there are no teaching professors. This would produce a safety net for some people. If such a safety net would exist it would be also be good to encourage people to do more high-risk research. If it then doesn't work they could still do teaching.

Hermann Bujard

I like to comment on that. I think you need a department system in which the faculty of the institute encourages young people to do novel research and not just to continue what you have done as a post-doc, and then protect them if they do not appear to have enough papers. Moreover, when we are sitting in DFG or other committees to evaluate, we should not count the papers. I am often shocked when I see colleagues who have not read key papers when evaluating someone. A young person who has not published much within three years and has no *Nature* paper, is hardly considered a human being worth to be supported! It is also part of our responsibility sitting in these evaluation committees to talk to the people and find out how good they really are independently of their number of papers. A department can shield young, talented people who after seven years or even longer, suddenly excel.

Walter Doerfler

I would like to respond to Jonathan Howard's comment on the lack of charity foundations in Germany. There are a few small ones in Germany, for example the Wilhelm Sander Stiftung that has supported our research on mechanisms of adenoviral oncogenesis. But there is no long-standing tradition in Germany to donate private money for research. I once approached an agency to support one of our projects. They considered it favourably, but finally decided that in Germany research was to be supported by the government and this was a set tradition.

For more recent consideration I wish to submit that, in Germany, severe damage has been done vis-à-vis the appreciation of molecular genetics by the anti-gene technology movement: mainly, but not exclusively, caused by the "green adventure". If you claim to make a contribution to cure cancer, you probably could convince someone in the wealthy community to donate money. On the other hand if you propose a molecular biology or genetics project and do not camouflage it carefully, you will probably not convince even people from Bayer. I remember a discussion we had at the Institute of Genetics with representatives from Bayer in Leverkusen. Herr Büchel told us on that occasion, "Well, you know your Institute is not so unlike Yale or Rockefeller, except, of course, they have five to ten times the number of professors you have. Moreover, you should remember that your science is not acceptable to society in Germany." This anti-science movement has done tremendous damage and had a major negative impact on molecular biology in the 1980s and 1990s in

Germany. While red gene technology now appears to opportunists in a rosier light, the wrath of the ideologues is targeted towards green gene technology. Altogether, this is the saddest experience I have had after returning to Germany in 1972.

Maria Leptin

There are no more urgent points now. There will be another chance to discuss things in the afternoon during the discussion on teaching. At half past two we will reassemble here for that discussion and for Peter Starlinger's concluding lecture. Thanks very much again to all the speakers.

VII. Establishment and Teaching of Molecular Biology in Germany

26

Panel Discussion

Charles N. David, Jonathan Howard, Hubert Kneser, Peter Overath, chair: *Benno Müller-Hill*

Charles David

I have been asked to say a few words about teaching and German universities. I will make it short and hopefully provoke a discussion. I am an American. I have lived extensively in both Germany and the United States. I have been in Munich for the last 22 years as an academic at the university. I also worked from 1968–73 at the MPI for Virus Research (now Developmental Biology) in Tübingen.

I find that the most conspicuous feature of German science is also one of the most conspicuous drawbacks of German science: that it is divided between the Max Planck Society and the universities. That is very bad for German science, at least in my view. The Max Planck Society is well funded and does excellent research but has no students, and thus no teaching. The universities are under-funded and do too much teaching, although there are some very conspicuous exceptions to this. Your Institute of Genetics in Cologne is one exception, the ZMBH (Zentrum für Molekularbiologie Heidelberg) in Heidelberg is another. Thus, although there are exceptions, by and large, German universities have a tough time doing internationally competitive research because they are drowning in teaching and do not have enough money. I think it would be very helpful if you could fuse the MPIs with the universities. In fact, that is the only hope for German science.

Let me give you an example of what could happen if you did that. The Max Planck Society is a competitor to German universities. Many very good German scientists end up at MPIs because the Max Planck Society has got the money and the name to attract them. That takes these people out of the university context and that is what is bad about it. In the United States it is exactly the opposite. There are no MPIs and there are only a few institutions similar to them, for example the NIH (National Institute of Health). Essentially, all good American scientists are professors in American universities. They see American students and American students see them; they have intimate contact with these professors. The reason I am in biology today is because, as an undergraduate at Harvard University, I had Jim Watson and Matt Meselson as teachers and worked in both their labs. If Watson and Meselson had been in Germany, they would likely be Max Planck directors and would not spend their time teaching undergraduate students. I think that is bad for German science and we should try to change it.

Jonathan Howard

When I first arrived here, eleven years ago, there was a huge barrier between professors and students. I discovered that there was actually a separate community of students because they had no real contact with the academic professors. They had a kind of student union structure, the *Fachschaft*, which provided all sorts of ancillary services that would normally be provided by the academic community. The *Fachschaft* still exists and fulfils a valuable role, but I think that this is an instance where one can see that there has been a definite mood change in Cologne at the Institute of Genetics, and I suspect also elsewhere. There is now the realisation that students are individually quite important and that they have a claim on the attention of academics.

It was a very unusual experience to arrive here and find that a student who wanted to work in one's lab could not bring a recommendation from an academic because nobody knew the student. That is clearly changing now, both generally and specifically with the rise of the mentoring system. Despite all the difficulties of the bachelors and masters degrees, the technical questions have brought with them a much closer attention to the needs of students and an improving relationship between academics and their students. The universities are certainly taking the students more seriously, at least more than they used to. The direction of that change is favourable and it will be spectacularly accelerated if the German universities introduce student fees. I doubt very much whether German university professors realise what they are getting into.

I want to go back to something which came up earlier in the meeting. That has to do with this idea of diffusion of knowledge. One of the points made repeatedly during this meeting was that German science and, in particular, German genetics was backward after the war — and the usual arguments were brought up. Indeed, looking at the German scene in the mid-1950s it seems that nobody was caring about the important advances being made in molecular genetics elsewhere. "Elsewhere" was the United States and Britain. And these were set up as examples of places where, indeed, huge progress had been made in a wholly new discipline. The typical German inferiority complex led to the belief that Germany was miles behind and had to be forced forward. And the result was the foundation of our wonderful institute.

I would like to tell you something relevant from my own experience, which has to do with the diffusion of knowledge. I was an undergraduate in Oxford from 1961 to 1964, so around the same time this institute was beginning to get going. Let us remind ourselves, 1961 to 1964 was a period which covered the decade (ending in 1963) after the publication of the structure of DNA. I was in England, 80 miles from Cambridge. I studied zoology. The main features of that study were, firstly, the evolution of animals, and secondly, genetics. I am sure you understand what is coming next. In the course of my three years we were taught nothing about DNA. The information that you were already well on the way to acquiring in Cologne was not available to students in Oxford. I want you to know that because you should realise that what is often perceived as a German problem may be a problem that touches everybody elsewhere as well. We were incredibly backward in Britain's second greatest university. We did not have any DNA teaching. Next time we talk about how backward Germany is, you should think about what I just said. Germany is very good at seeing these disasters, drawing the wrong conclusions, and then making the right reactions — in the present case, the foundation of the Institute of Genetics in Cologne.

Let us come to the present and the future. The discussion we had this morning (panel discussion 1) was full of interest but there was an uncertainty at the base of it, of whether we were talking about a research institution or a teaching institution. I discussed teaching and now we have to find out whether it is indeed possible, to come back to Charles David's point, to have an effective teaching institution which is also doing world-class research. I do not think that the Max Planck system has necessarily to fuse with the university system to rescue German science, but maybe it should come a lot closer. The American system that Klaus Rajewsky highlighted in his contribution has a mechanism whereby research achievements benefit universities through the payment of indirect costs. In the United States, universities become richer in just that sense in which universities wish to become richer, because the quality of the research that is being done in them improves. They are able to afford better resources for everything because better research is being done.

We do not have an indirect cost system in Germany. What we have here is what we in Britain would call the dual support system, whereby the government provides the universities with the resources to make what is called the well-founded laboratory and the infrastructure of teaching. The marginal costs of experiments are supported by the *Deutsche Forschungsgemeinschaft* (DFG). Everyone who has ever had a grant from the DFG will know that there is not enough money in it to set up a lab. You only have enough money to run something which is already in working order. And, as we all know, the amount of money that comes from the Land or the government is dropping all the time. It is getting harder and harder to establish a well-founded laboratory in the present climate. It would not be wise merely to say that we needed more resources for science, because of course we need more resources. But I think we can run a first rate scientific system here if we reward good science by rewarding the universities in which it is done. Therefore, I would like to see a shift in resources towards the DFG and the provision of indirect costs. The Max Planck system, for all its grandeur, is undoubtedly over the top. It is not difficult to find *Abteilungen* with 20 or 30 fully-funded staff — really huge. One would like to see a better balance. But I would like to see much of the money going to the DFG to fund an indirect cost structure. That will mean that we can run a research environment in the universities and teach students in a first rate research context.

Anyway, there are many prospects for the future, a little bit of redirection of resources and putting more money into reward of research.

Hubert Kneser

I will tell something about my personal experiences in teaching. I was trained as a physical chemist and I had the opportunity to gain insight into the new molecular biology very early. For example, I had the lucky circumstance that I heard about the Hershey-Chase experiment even before it was published. Hans Friedrich-Freksa, the director of the MPI für Virusforschung in Tübingen, with whom I had a lecture course, told me about it during a subsequent examination. So the diffusion of knowledge from a MPI to a university happened in some lucky places. During my first job at the MPI in Tübingen I was told that there was a phage course in Cologne where I should go and learn all about the new molecular biology. When I was in Cologne, my first job was an assistantship with Peter Starlinger. I slipped into molecular biology and the genetics pathway from a different direction.

There was no big teaching load in Cologne back then. My first teaching experiences were in the phage course, where we were amongst us just like in a family. Everyone knew everyone else and the interest was very strong. It was more of a discussion than teaching.

Since biology students in Cologne did not have to take genetics courses there were many professors at our institute, but only few students. We usually organised special courses with two professors, e.g. with Peter Starlinger. We taught virus genetics, radiation biology and other topics and we had about five students.

Then I took over the teaching of medical students. They were taught in a completely different atmosphere. Medical students were not primarily interested in the beauty of DNA and the elegance of molecular biological mechanisms. What interested me did not have a lot to do with what they wanted: to cure patients of diseases. In fact, this attitude is partly right because the machineries of DNA replication, transcription and translation are simple and clear, but also stupid. The polymerases not only read the DNA exactly but they copy good genes and junk DNA equally well and the ribosomes do not understand what their proteins will do. This stupid machinery has no direct appeal to medical students.

I think these mechanistically beautiful things that were elucidated in the 1960s have been falsely received by the broad public. Many people thought that genetics was simple and that geneticists now knew everything. That led to the false impression of *der gläserne Mensch.* But these simple fundamental mechanisms were just a bottleneck of evolution. If evolution wanted to create and replicate complicated organisms, it had to invent a machinery which did it just mechanically, like copying a tape. Everything else, from proteins to the immune system, is utterly complex. This false, simplified view is not confined to medical students or the general public without knowledge of molecular biology. Carsten Bresch told me in the 1960s that once we had elucidated the mechanism of recombination we would understand *E. coli* completely. A few years later there were two thick books about the lac operon and the phage lambda which were complicated, but not nearly as complicated as *E. coli.*

Genetics generally deals with apparently simple mechanisms. That happens since Mendel's discovery of the one-to-one segregation and the chance recombination of chromosomes, and Darwin's proposal of simple selection of the fittest and its defined simplifications in social Darwinism, which had fundamental consequences, especially in Germany, but also other countries. As Jim Watson said about some human properties: "It is all in the genes." This seems to me to lose sight of the complexities which surround these simple mechanisms.

In the discussions among medical students in the 1970s one urgent problem was whether abortion after prenatal diagnosis should be made possible. I remember very well the change in the consensus on how to handle the problem. In the 1970s it was a broad consensus that the physician is the one who has got experience and has to give advice to the patients. Twenty years later it became the consensus to let the parents or the family decide what to do; to just give them the necessary information and be careful that the information arrives at the right place.

The gap between the simple molecular mechanisms and the phenotypical complexities is well exemplified, for instance, with cases of cystic fibrosis in a family where we know that two brothers have exactly the same alleles of the gene. Nevertheless, they do have different symptoms: one in the lung and the other one in the pancreas. So other genes have an influence on the phenotype, too.

We have to sensitise the growing physician to these complexities and should never rely on the simple mechanisms they have in mind when they say that "it is all in the genes". The interplay between nature and nurture, between genetics and environment is complicated and that is what I wanted to tell these students. Bridging the gap between our well defined knowledge in molecular biology and the vagaries of practical medicine is very fundamental.

Peter Overath

I think Charles David and Jonathan antedated the two points I wanted to make. I have done very little teaching in my life. I was here at the Institute of Genetics from 1966 to about 1973, and at that time we had to do very little teaching. Afterwards I had the privilege, however you may consider this, to be at the MPI in Tübingen for 30 years. Now I am back at the University, and there I again do not have to teach because I am beyond the age.

The role of Max Planck Institutes in the development of a local university is, of course, a very important point. First of all, I do not think that the idea of the fusion is a particularly good one or know whether it would work. After all, the MPIs together get about as much money as two big universities. But on the other hand, my own experience in Tübingen in trying to teach at the university was partly positive and partly rather negative. When we gave two-week courses at the MPI for students, with lectures, experiments and discussions, this worked out very well. But I remember that at the end, when I was trying to give a rather elaborate lecture on the molecular biology of tropical parasites, I had only three students sitting there. Each time I went there I was hoping that these students would turn up. This is probably not the way to integrate Max Planck people into the university system. As I hear now that the whole system at the university is going to change because of the master degree and so on, one should consider including the Max Planck people on all levels, from the directors to the research assistants, into the regular teaching curriculum of the students doing their graduate studies. This would probably help the university staff to reduce their teaching load, and it would be good for the Max Planck people to give their knowledge and experience and, of course, have contact with students.

The second point is the indirect costs. I think the only way that this can be solved is if the universities are better supported. And for the support of the universities, the indirect resources which Jonathan mentioned are important. Because then the university really has the interest to get the best professors and students, because they get more grants and therefore more indirect resources. This would, of course, mean that there will be fewer grants, unless the total amount of money from the DFG is increased by about 50%.

Jonathan Howard

I have learned that there can be bad relations between MPIs and their local universities, but I have not learned that from my experience in Cologne. I want to say that with the utmost emphasis. Although

I know that there have been rocky relations between the university and the MPI für Züchtungsforschung in Cologne-Vogelsang in the past, at least during the 11 years that I have been here, the relations have been good and increasingly excellent. There has been a massive amount of participation in teaching over these 11 years, at all levels, from *Assistenten* to director. It is an extremely positive relationship which can only enhance the value of the Cologne community and the attraction of the Cologne community to students who are wondering where to go to learn some good science. It is very important to stress that bad and uncomfortable relations between MPIs and universities are not always the case in Germany.

One of the problems that the Max Planck system has is that it seems, somehow on purpose, to put its institutes as far as possible from university installations. They are often on a hill where the university is in a valley, or vice versa, and they are unreachable by public transport. So they become very alien places. I think that is a terrible mistake. If I were starting the Max Planck Society from scratch, I would put all the MPIs in universities. What we want to have are big, powerful institutes on the Max Planck side, closely embedded in university environments. Because students need contact with high quality research, there is no question about that.

One of the problems of funding universities in relation to the quality of their research — which I am, in principle, in favour of — is that those universities which do not support very active or productive research communities fall away in functionality as teaching universities because the students are not given contact with high quality research as a part of the normal daily routine. This is something which is affecting England very severely at the moment. They have adopted what they believe to be the American system (it is not, in fact, at all the American system) of reviewing science all the time or reviewing universities all the time, and giving departments scores. Those departments which get a score below sort the highest level basically have money taken away while the ones that have the highest scores have more money given to them. The criteria we used were: how many papers are published; in what journals have they published; how many Ph.D.s; how many honours are awarded; and so on. If you do this you get a fine, positive feedback and great universities become greater but lesser universities become weaker until, in my view, they cease to function as a university at all, but remain rather depressing teaching institutions. That is a risk, and one has to be careful to avoid it.

I greatly respect the German intention to maintain a certain level of quality in universities. I think that is very fine. But I am a bit worried about this elite stuff and fear that we may be going the same way as Britain. When you visit Oxford and Cambridge you think, "my goodness me, the British academic system is really wonderful", but if you go to most other places you will not feel that at all. And that is the risk: that German authorities are not close enough to the business of science and science teaching, and are perhaps not adequately aware of the down-side of the elite university idea.

Charles David

I see we have precipitated something with this discussion of Max Planck and the university. I quite agree with you that the relationship between MPIs and the universities is getting better with time. Indeed, there are now spectacular examples of a close interaction; for example, in Cologne between the university and the MPI in Cologne-Vogelsang. I think it is very helpful to put MPIs next to universities, and this is exactly what we have now done in Munich. We have moved the biology department and also the gene centre right next to the MPIs for Biochemistry and Neurobiology in Martinsried. These are very positive steps and, if they continue, I hope that we can draw more Max Planck people into universities and connect more students with Max Planck labs. It is all about getting people together. I think it is worth mentioning, in this context, one other example. Twenty years ago an analogue to the Max Planck Institutes was created in the United States. They are called Howard Hughes Medical Institutes, and they were established within universities. That is one of the major reasons why I would like to move Max Planck Institutes not only closer to, but into the universities.

Peter Overath

There are many reasons why the Max Planck Society put certain institutes at certain places, and whether or not they are integrated into the university directly on the campus. I think there has never been an intention to put the institutes outside the universities on purpose. Having studied chemistry at the Universität München in institutes located in the centre of town, I was very sceptical when I first heard that the Max Planck Gesellschaft planned to built their new institutes in Martinsried. In hindsight this was a very wise decision, because eventually the university institutes followed, and, including the biotech companies, the Martinsried complex is now a unique place of excellence in this country.

Charles David

Perhaps, in this context, it is worth mentioning another point: the Institute of Genetics was founded at the university because Max Delbrück wanted to establish it in a university setting. He was, as far as I know, not interested in having it associated with the Max Planck Society. I would say that this was very positive for the German universities.

Hermann Bujard

I do not know how to ban this discussion, but for me it sounds like what we are discussing right now is not the main problem of universities. Our university's problem is that we are vastly underfinanced. Even if we have some institutes which look better, like the one here, and ours in Heidelberg, in general we have a highly underfinanced university situation. At the same time a huge amount of money is spent for research outside the university. I mentioned today the Leibniz and the Helmholtz-Gesellschaften. Some of these big institutions are nothing more than platforms, and if you ask them what they do as a platform they do not have any idea. Now that is a bit polemic. We have the so-called "*Spartenforschung*". Every minister has some research being done somewhere. This is a huge amount of money in Germany and it is not reviewed. When we think about our resources and about the importance of the university as the prime institution where our young generation is being educated, then they are totally underfunded, just like some of our other educational institutions. This should be the problem to talk about, and not where Max Planck and the universities are situated. This is only a secondary problem.

Joseph Lengeler

May I just mention another major problem: at the universities in Germany we are confronted with another problem. We are overcrowded with lousy students. Eighty percent of the students we have in Osnabrück should never have been at the university, they should have been at what is called the *Fachhochschule* in Germany. They want this kind of practical education and they would be highly motivated for these kinds of studies. They were not gifted enough for the university and basic research, and were not really interested. Most of our time is spent teaching these students and we thereby dilute the good teaching. That is a major problem.

A Ph.D. student

I am a Ph.D. student at a research school and for me it is also a question of motivation. I think, in general, we can get a more-or-less good education at the university the way it is now. But in the end the problem is that we have many students, at least in biology right now, and mostly they just are not motivated. I think if you integrate the MPIs into the universities you will see how high the quality can be, because the quality of research in an institute is much better. I myself would never go to a university like Kiel because it is simply not well recognised. So, for my career, of course I would go to a Max Planck research school just because the recognition is better. In the United States you have the same thing, for example, what you said about the Howard Hughes students at university. In the United States, on the one hand it is fortunate that the Howard Hughes Medical Institute is integrated into the university, but, of course, I would pick an elite university or an elite institution. In Germany the MPIs and the other outside research institutes are exactly the same things, I do not know what the big difference is.

Another Ph.D. student

I have got the impression that society as a whole is becoming more and more uneducated. The reason for this is the schools and the teachers. And if you have a look at who is responsible for teaching teachers, this is the university. This is a big problem in Germany.

Maria Leptin

This is a statement which is often heard. Yes, society seems to be less educated and we like to blame the schools, and the schools are in desperate need of improvement. But Germany and America are very rich societies. The other point is that we do have too many students in biology. We do not need 360 biologists in every major university. Just add it up! What is going to happen to them? We are rich countries and cannot just say, "Let them go unto the soil," or, "Let them go and be bank clerks," or whatever. I think rich countries should afford to train a very large proportion of young people for as long as possible. And that is what drives politicians to send us more and more students, much to our dismay. This has caused a despair among universities that has not really been addressed at all. We should train more people, but we cannot do so because we are overstretched. Politicians seem to think that it works by just stuffing more and more into universities. The trends in the 1960s that were supposed to be partly temporary have never been reversed and have never been cured. We all agree what is bad and we all know where it comes from, namely from the politicians. But again I have do ask, "What do we do about it?" It must be changed. High level teaching quality suffers and solutions have to be found. I do not have any but I would like to hear about ideas.

Charles David

Maria is quite correct. Solutions have to be found. German universities have a very serious problem and it is getting dramatically worse. There is, however, at least one solution on the horizon. Probably it will not be very popular, but it is high time that German universities take tuition fees for what they do. This is probably going to happen in Bavaria this coming year: tuition fees will be a thousand euros per student per year. In comparison to American and English universities this is peanuts. Nevertheless, people are complaining about a thousand euros per student per year. Do you realise what tuition fees would do in the way of solving some of our financial problems? The discretionary budget of the University of Munich, which is a relatively big university, is about 40 million euros a year. That's all. If every student of the University of Munich paid a thousand Euros in tuition, that would be 40–50 million euros and would more than double our budget. This will be a fantastic improvement. The argument against tuition, which I do not subscribe to, is that it is unfair to people from poor family backgrounds. This is true, but it is also not relevant, because students who go to German universities tend to come from better financial backgrounds. And for these people, who have more money anyhow, to argue that they should get their education for free, I consider an outrage.

Gerrit Praefcke

My question is about people teaching in the United States. Not everyone studying biology in the US is at the Howard Hughes

Institute or Stanford. So what happens to people who study biology in North Dakota? Or, are there none? And are they taught better than people at an average German university?

Jonathan Howard

I do not know whether the situation has changed now, but when I used to spend lots of time in the United States I observed a very interesting phenomenon. I was always in what one may call an elite university, but the places where my colleagues had been undergraduates were all over the place, and certainly mostly not elite institutions. Many of them very good, but they were not research but rather teaching institutions. These institutions, places like Bryn Mawr, Smith, Wellesley and so on, were very much focussed on teaching. It seems as if the scientists who made it into the faculties of the great universities had gained their knowledge and motivation during their undergraduate time all over the place. And then was the question, where did they go for their Ph.D.? They moved into the best competitive institutions, whereas German students tend not to move for their Ph.D. There is a very strong tradition here to stay put. It was a very striking phenomenon in the US, and it was a different atmosphere altogether. And the undergraduate stage was clearly seen as being quite different from the research stage, with its own priorities.

Speaker from the audience

I have got a few questions. First of all, what is the purpose of teaching? Is it just training new scientists and academics or is it also for the industry? Because there are not many links between academic life and the industry. Or is it teaching to have new teachers? And is the only way of having an academic career to become a professor, which seems to be like a dark tunnel where you have to go a very long way, and at the end of the tunnel there is some light. But is there anything in between? We used to have a *Mittelbau*, which included the duty of teaching, but I have the feeling that it does not exist any more. There are many professors who are only interested in doing research, and teaching is not something that is a favourite thing to do. So the question is: what we are teaching for and who should do it, the professors or the post-docs? The post-docs only have positions for two or three years. The only way you have a future is when you publish, but when you teach it is not acknowledged.

Charles David

I am not sure whether you were speaking to me but I have a strong opinion on this. Why teach? Teaching is fun. I like doing it. That is a good reason to do it. The second reason to do it is that raising the level of knowledge of anybody is good. Finally, I do not teach students so that they can go out and work for a bio-tech firm. I am there to teach students about biology. If students can take that knowledge and work for a bio-tech firm, that is fine. But I do not think universities should be vocational schools for the bio-tech industry.

Speaker from the audience

I want to say one more thing. If you want to do science there are only a very limited number of positions where you can do it. I sometimes have the feeling that there is no life after a post-doc. There is the post-doc and then you are a professor. Recently, there was an article in the *Frankfurter Allgemeine Zeitung* which said that they want to have more women in research, but it is very difficult to find positions where you can do it.

Maria Leptin

I think one might say there is indeed no other way. You do a post-doc and then you become professor and you become independent. That is how ideally it should be. Our brains are best when we are young. So people who have proved during their Ph.D. and their post-doc time that they are good, original, and that they enjoy doing science should then be given complete independence. That is how it works in the United States. I think it is a problem in the German system that we have this *Mittelbau* idea as something sacred. That is a haven for people who do not want to or cannot become professors. We always look at America and their many problems, but the early independence is actually something we should strive for. If we did that and everyone had independence right away, then we would solve several problems. We wouldn't have huge hierarchical structures where the professor at the top is never seen by the students and we would solve the problem of jobs for younger people. We would not have *Assistenten*, but we would have a direct transition into independence and full responsibility for everything.

A Ph.D. student

I strongly disagree with scientists from MPIs being involved in teaching, because there is no necessary connection between being a good scientist and being a good teacher. A good scientist is not always a good speaker or teacher. To motivate students you need a good teacher and you need to make the students think about problems. The professors at universities should do the same thing: make students think more about various problems in biology. This in turn helps their Ph.D. students because at some points the Ph.D. students do not get enough guidance and time for discussion with the professors. Another thing is that German students are more reluctant to ask questions, even if they might not be stupid questions. The last point I want to mention is the exposure of students so that they get to know scientists in the MPIs or other top scientists. I come from India and we have a programme called the "summer research programme" whose purpose it is to get to know people in research. And the kind of research going on in the university becomes clear to students when they are doing their diploma or master programme.

Benno Müller-Hill

I would like to comment on one point which you mentioned, concerning students asking questions. This was different in the 1970s when I gave the general lecture here. There were questions for half an hour or even longer, so that we had extra hours for questions. This has now stopped almost completely. I would say, what also stopped almost completely is that students come to seminars or lectures which are not involved in their exam or which do not give them any direct advantage.

Speaker from the audience

I think the motivation has to be on both sides. If you want to motivate students to learn you have to have teachers who are motivated to teach. Of course there are always people who are not interested or lazy, but I think the biggest proportion of the students do biology because they are interested. I also think that some of the professors achieve a spark of interest and relate the fascination for the subject, and you can see that people love to teach when their lectures are always full.

Charles David

The comment that was made before about the poor correlation between good teachers and good researchers, I think is not true. Superficially, perhaps, it appears true; there are some good scientists who are poor teachers. A closer look, however, suggests this is not the case. Let me give you one example: Jim Watson. At first glance he was a lousy teacher: he mumbled, he talked to the blackboard and he was rather shy. However, his course on the biology of viruses totally fascinated me when I was an undergraduate. It was fantastic. That is the reason why I subscribe to the idea that the best researchers are frequently, almost generally, the best teachers.

Jonathan Howard

I was never taught by Jim Watson, but I was taught by Rodney R. Porter. I suspect that he was probably technically a worse teacher than Jim Watson. Still, it was stimulating because the intellectual standard was high. Porter got a Nobel Prize for the structure of antibody molecules. The definition of a good teacher is hopeless. If you start doing teaching exams you do not know what is going to spark people. Teachers have to say something interesting, but it is up to you to respond.

Joseph Lengeler

An excellent scientist may be a weak teacher, but a lousy scientist is always a lousy teacher. That is rule number one. If we encourage young people to study at universities in huge numbers, we should tell them that if they have short legs they should not enter high jump contests. They should then look for the area in which they are highly fit. That is because motivation alone is simply not sufficient. The third comment is about mentioning that the *Mittelbau* is not a solution. In some way I do agree, but in Germany it was a solution for a long period of time. Now we have ended this solution but we didn't find an alternative, and that must be fairly mentioned. In the United States you have this drop down phenomenon as people who did not succeed at Harvard become professors at a lower university, in colleges and schools.

Jonathan Howard

A word here, because we are talking about the history of the Institute of Genetics, where the *Mittelbau* still exists and is still functional. It contributes a huge amount not only to teaching, but also to the organisation of teaching at the Institute. They have probably been, at least at the time I have been in the Institute, pioneers in organising teaching for teachers. There was a disparaging remark made about the quality of teaching that teachers are getting and I think this is not appropriate. As an English person I am hugely impressed by the quality of teaching the teachers are getting here. It is quite in a different league from what is being done in Britain. So there is still a *Mittelbau* and this Institute uses it in a very creative way to provide a remarkable backbone structure to the Institute — something which in an *Abteilung*-based system would be very hard to provide. So just to be clear on these two points, they do exist here and they do a fantastic job.

VIII. Final Remarks

27

Science and Society

Peter Starlinger

The organisers have done very well to include the topic "Science and Society" in the programme. Our society is nowadays very interested in how science justifies what it does. This must be so, because it is society which provides the considerable sums that are needed today to subsidise scientific research, and also because the results of scientific research have an ever increasing impact on society itself.

Genetics has not done particularly well in terms of public relations. A considerable segment of the public is highly critical of, or outright hostile to, genetic research in general and to its applications in agriculture, in particular. Others tell us, sometimes a little condescendingly, that this is our own fault. They say that we do far too little to communicate with society as a whole and thus we should not be surprised that the public listens to Greenpeace and forms their opinions accordingly. We should leave our ivory tower. This, however, is more easily said than done. To communicate effectively with society requires television time or access to major newspapers, and this is not obtained easily. Journalists set their own agenda and are not keen to have outsiders express their views. Thus, it is required to have either good luck or well developed communication skills in order to exploit the opportunity to address a larger audience on scientific matters.

Let me try to say what we have done over the years. It was little enough, but it was the best we could do. One activity was the initiation of the Spring Meetings, which grew out of the phage and bacterial genetics courses. These courses, modelled after Max Delbrück's famous phage course in Cold Spring Harbor, and started on his advice, were held in Cologne throughout the 1960s. The courses were well received and attended by quite a number of colleagues who later held important positions in universities or institutes of the Max Planck Society. Thus, Max Delbrück's wish to disseminate both modern molecular biology and the easy-going style he had introduced in Cologne was served by these courses.

Following this tradition, we invited scientists to give seminars at these courses. We quickly realised that some of our students, particularly those with a less than satisfactorily developed background understanding, would be unable to grasp complicated seminars presented by first-rate speakers in the first week of the course. In addition, some of the most important people would not be readily prepared to come to Cologne following Max Delbrück's departure. We therefore had the idea to organise a meeting at the end of the course, once students were well prepared, anticipating the likelihood of one guest speaker luring another to attend on the basis of his credentials and the newly won receptiveness of students. The first of these courses was held in 1964, and our idea worked out quite well.

There is another point that I wish to make. These meetings were freely accessible. It was not necessary to provide a CV and a list of papers in order to prove that you were really interested in this field. While we did not provide anything for the attendees, who could sleep in a tent if they so desired, there was no charge for the meeting and no limits imposed in the number of people admitted.

Of course we have to thank the *Deutsche Forschungsgemeinschaft,* which provided us with the necessary travel expenses that, used thriftily, would bring the speakers to Cologne without any fees required of the participants. Over the years, particularly after new professors, such as Benno Müller-Hill, Klaus Rajewsky and Walter Doerfler, had come to Cologne and had done much to broaden the scope of these meetings, these became quite popular with graduate students and poorly paid post-docs from all over Europe who would otherwise have had a hard time gaining admittance to prestigious meetings in the Swiss Alps.

While these meetings were mainly an internal scientific affair, we thought we should try to reach a wider public. We held the view that school teachers should know something about modern genetics. When we started this in the 1960s, there existed none of the current hostility toward the science of genetics. But, of course, many of those school teachers had received their training much further back in time, even before the second world war. They had not yet learned anything about modern genetics. Therefore, we organised courses for them with the help of Kurt Schlösser from the MNU, the *Landesstelle für den Mathematisch-Naturwissenschaftlichen Unterricht*, a teacher-training institute administered by North Rhine-Westphalia. We had teachers coming to our institute for one day or, sometimes, for two days, and we were discussing with them what we considered the wonderful world of modern genetics. In addition, we were also talking to them about ways in which this knowledge could be used within schools and how we could help them to undertake demonstrations and experiments in their classes, which were bereft of fancy equipment. Could something be done with very simple kitchen devices? Some of them got enthusiastic about it and had very good ideas. Where we had a surfeit of old and unused equipment, we donated it to the teachers (please don't tell the *Landesrechnungshof*!). Over the years, around 200 to 250 school teachers participated in these courses. This was a significant percentage of the biology teachers employed in the state of North Rhine-Westphalia at that time.

We once had a particularly interesting experience. In the late 1960s, many young people in high school chose not to study the subjects of Latin and classic Greek. Therefore, some of the teachers of these subjects became "surplus to requirement". The province of North Rhine-Westphalia started a programme to quickly retrain them in courses for different teaching facilities, and some of them expressed a desire to study biology. What I am going to tell you now relates to something that was done not by the Genetics Institute alone, but by the whole of the Cologne Biology department. In our institute, we were responsible for molecular biology. The participants were very interested, and from them we learned some philosophy — they were quite interesting people. Some of them said that they had originally considered studying biology, but had finally settled for Greek literature, Socrates and Plato. It was a very stimulating time we had with them.

By the 1970s this kind of teaching had died out — partly because Kurt Schlösser left the MNU and went to work at the university, partly because the state could no longer afford to attract teachers to Cologne from elsewhere, and partly because the influx of university students grew so much that there was neither space nor time to run these courses. They were taken up again only much later, in the late 1980s and the early 1990s, when we again began to undertake teacher training. The first step was taken by Heinz Saedler, who had experienced much trouble with his work on gene technology and the first German experiments with transgenic plants outside of the greenhouse. In order to show the public that transgenic plants do not bite and have considerable advantages in the right circumstances, he built a demonstration garden together with Wolfgang Schuchert, affording school classes and other citizens the opportunity to come and see for themselves and obtain first-hand information.

Another activity was provided by the Institute of Genetics, mainly at the behest of Heidi Fußwinkel and Klaus Reiners. We had to visit different cities to meet teachers, simply because there was no money to bring these people to Cologne. So we thought we should visit them instead and provide them with lecture courses and sometimes a few demonstrations. Both of these activities were joined in a very genial institution, the Köln PuB, the name of which should not give you the impression of Kölsch or Guinness beer, but which is an acronym for *Publikum und Biologie*. This is a club that has the intention to interest and to educate the public. Köln PuB now has its own laboratory, and they once again provide courses for teachers. I think this is a very good and promising development.

What else did we do? I remember from the 1970s the first *Ringvorlesung*. This is a series of lectures given by different lecturers, each of them coming up with one particular topic. The first *Ringvorlesung* was on evolution. It was open to the public and it was announced in the newspaper. Many people came from the city to attend on any weekday evening. It was quite interesting for us and there was interest on the part of the audience too. We repeated this the following year, trying to organise a *Ringvorlesung* on the topic of information. However, this proved less of a success because fewer people came, and it died on the vine. I will discuss another *Ring-vorlesung* towards the end of my talk. But before that I must, of course, mention discussions about gene technology.

Starting with the Asilomar conference in 1975, there was enormous public interest in gene technology. Gene technology quickly gained a very bad reputation, but not all people were convinced that this new development represented an evil, and they wanted to know more about it. Thus, we were often invited to give lectures and hold seminars for very diverse groups: trade unions, employers, church groups, and even political parties. I think I can say that whenever such an invitation was received, it was accepted. But how much does that count? If you go to a certain group one evening, you may have 15 to 20 people or, if you are more successful, even 40 people. But in a city like Cologne, with its population of one million, this is not really very much. However, we did what we could.

All of this was somehow related to genetics and it was natural for geneticists to engage in such activities. Are we allowed, however, to talk about problems that do not fall within the domain of our own specialty, living as we do in a university environment? An important contribution of the Genetics Institute was in the form of Benno Müller-Hill's historical studies on crimes committed during the Nazi era. These are now generally well acknowledged. In the early days, however, it took quite some courage on the part of Benno, because many people did not want to hear about this.

Another non-genetic, extra-curricular activity undertaken by members of the institute was mentioned yesterday by Carsten Bresch. It happened at the time of the "Spiegel affair", a huge scandal in Germany. The Federal Minister of Defence, Franz-Josef Strauß, had the publisher of the well-known German news magazine *Der Spiegel* thrown into prison on charges of high treason. The staff of the institute began collecting signatures and sent a letter to the Minister of Justice asking for the release of Mr. Augstein. The letter was very polite, in keeping with the times prior to 1968, yet it led to some controversy over the issue of whether scientists should engage in such activities.

The 1960s and 1970s became more and more politicised and, to a certain extent, this culminated at the end of the 1970s and the beginning of the 1980s in the big peace movement, which grew out of the increasing concern of many people in the country that the ongoing arms race might end in a catastrophe. This is hardly imaginable today. In those years, while it was officially said that the arms race was only to deter the adversary from starting a war, many people thought that this might go wrong. During the same period as this growing peace movement, in which many people and also many scientists were involved in different ways, we in Cologne once again had a *Ringvorlesung*. From our institute, Hubert Kneser and I were involved. Then there was the late Christiane Rajewsky, Klaus Rajewsky's first wife, who was a professor of politics, and the late Ulrich Klug, one of the fathers of the reformation of German criminal law, a much acknowledged law professor and also a politician (at one time he was the Minister of Justice in Hamburg). Also involved was Karl Bonhoeffer, the son of Karl Friedrich Bonhoeffer of the Karl Friedrich Bonhoeffer Institut in Göttingen, of whom you heard yesterday. He was a professor at the medical school. And there were a few others. So we organised a *Ringvorlesung* for six years, that is, 12 semesters in all. Every Monday in the semester we had this *Ringvorlesung*, and we had a lot of guests. In addition to the people from Cologne, we had many guests from outside, among them many ministers, politicians, writers and actors.

We started holding these sessions in the biggest lecture hall of the university. This was a little bit difficult, because we had to testify that these lectures were within our realm of teaching. Otherwise, we would not have been allowed to use the big lecture hall. However, when Ulrich Klug confirmed that this was a public teaching activity, he could not be refused. In the beginning, we filled this big hall with more than a thousand people. This could not be sustained later, when fewer and fewer people attended, leaving us, towards the end, with a regular audience of between only 30 to 60 people. For the longer period of six years, every Monday during the semester, however, I think this was not really bad.

Now you will ask whether all of this was to any avail. At first sight the answer has to be in the negative, because, as is obvious to us today, neither the politics of the Federal Republic regarding armaments, nor the attitudes of the public towards gene technology were changed by these undertakings. However, I still think that it was useful to do something like this. Why? Because sometimes, through mechanisms which we do not understand, the mood of the public changes. You cannot predict whether this will be a slow process or a rapid one. You do not know when it will occur. But I am pretty sure that if there is no public discussion of alternative ways of thinking, however small, there is no prospect of a change in the perceptions of the general public. Let us hope that, in spite of this widespread rejection of gene technology by the public, things are still occasionally subject to change. When Vesalius first dissected human bodies in the 16th century he had to take shelter at the court of the emperor in order to avoid persecution as a heretic. Nowadays, anatomy is a normal part of the preclinical curriculum. Let us hope that something like this may also occur one day in regard to gene technology.

Let us hope that our successors here at the Institute will also find the time to participate in the public debate, even if it sometimes seems to be a little tedious and time-consuming. Then, on the occasion of the 100th anniversary of the Institute, your successors will hopefully live in a peaceful environment without an arms race, where their science is discussed critically, as it should be, but where the right to practice science and pass on its benefits to mankind are acknowledged and appreciated by the general public.

Pictures of the History Workshop in Cologne in April 2005

1. Siegfried Roth
2.
3. Nagarajavel Vivekananthan
4.
5.
6. Gertrud von Hesberg
7. Hannelore Wirges-Koch
8. Cemalettin Bekpen
9. Christoph Rohde
10. John Collins
11. Andreas Paukner
12. Hubert Kneser
13. Rainer Hertel
14. Helmut Klein
15. Maria Leptin
16. Joern Puetz
17. Wolfgang Michalke
18. Hildegard Michalke
19. Gert Wuesthoff
20. Petra Pfeiffer
21. Elke Rottländer ?
22. Asa Boehm
23. Peter Overath
24. Joseph Lengeler
25. Boerries Kemper
26. Ruth Ehring
27. Bhupendra V. Sharvage
28. Peter Starlinger
29. Edith Tillmann
30. Thomas Trautner
31. Georg Michaelis
32. Kevin Johnson
33. Horst Feldmann
34. Paul Habermann
35. Ariane Toussaint
36. Walter Doerfler
37. Hans Bremer?
38. Simone Wenkel
39. Heinz Saedler
40. Frank Sprenger
41. Charles David
42. Benno Müller-Hill
43. Jens Brüning
44. Diethard Tautz
45. Teresa Corona
46. Hermann Bujard
47. Matthias Cramer
48. Klaus Rajewsky

49. Jürgen Dohmen
50. Volker Schirrmacher
51. Alfred Sippel
52. Gerd Hombrecher
53. Jonathan Howard
54. Bernhard Mühlschlegel
55. Ute Deichmann
56. Bettina Montazem
57. Madhusudan
58. Klaus Reiners
59.
60. Steffi Könen-Waisman
61. Gaby Vopper
62. Rita Lange
63. Parisa Kakanj
64. Thorsten Buch
65. Brigitte von Wilcken-Bergmann
66. Stefan Weisshaar
67. Natasa Papic
68. Nina Schroeder
69. Sigrun Korsching
70. Julia Hunn
71.
72. Yang Zhao
73. Gerrit Praefcke
74. Robert Finking
75. Michael Knittler
76. Sascha Martens

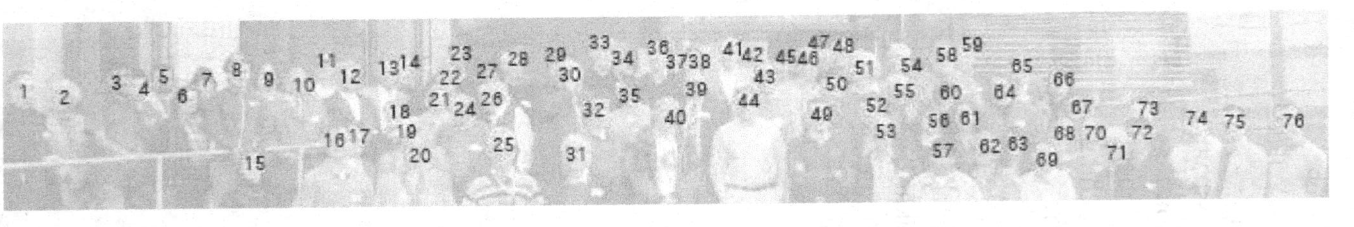

Fig. 27. The participants of the History workshop in Cologne in April 2005.

Fig. 28. See Fig. 27.

Fig. 29. Conference dinner at the new building of the Institute of Genetics. From left to right: Cemal Bekpen, unknown, Bernhard Mühlschlegel, Ariane Toussaint, unknown, Ute Deichmann, Bruno Strasser, unknown and Simone Wenkel.

Fig. 30. Ute Deichmann and Fritz Melchers.

Fig. 31. The participants of the panel discussion about molecular biology and the German university structure. From left to right: Hermann Bujard, Walter Doerfler, Maria Leptin and Klaus Rajewsky.

Fig. 32. Heinz Saedler.

Fig. 33. The participants of the panel discussion about the establishment and teaching of molecular biology in Germany. From left to right: Charles N. David, Jonathan Howard, Benno Müller-Hill, Hubert Kneser and Peter Overath.

Fig. 34. Walter Doerfler.

Fig. 35. From left to right: Maria Leptin, Carsten Bresch and Peter Starlinger.

Fig. 36. From left to right: Thomas A. Trautner and Bernhard Mühlschlegel.

Fig. 37. From left to right: Benno Müller-Hill, Lothar Jaenicke, Hermann Bujard and Ariane Toussaint.

Fig. 38. From left to right: Matthias Cramer and Peter Overath.

Fig. 39. From left to right: Klaus Rajewsky and Sigrun Korsching.

Fig. 40. From left to right: Hans Bremer and Hermann Bujard.

Fig. 41. From left to right: Peter Starlinger, Thomas A. Trautner and Carsten Bresch.

Fig. 42. Rainer Hertel.

Notes on the Contributors

Niels Bohr[82] (1885–1962) studied physics at Copenhagen University and completed his Ph.D. degree in theoretical physics using the electron theory to describe the properties of metals, with C. Christiansen, in 1911. He spent his post-doctoral years with Sir J. J. Thompson at the Cavendish Laboratory in Cambridge and then with Ernest Rutherford in Manchester. He worked on a model of the atomic structure which is still used to describe the properties of different elements. During 1911–16 he held lectureships in Copenhagen and Manchester; in 1916 he was appointed professor for theoretical physics at Copenhagen. In 1920 he was appointed head of the university's Institute for Theoretical Physics, a post specially created for him. In 1922 he received the Nobel Prize in Physics "for his services in the investigation of the structure of atoms and of the radiation emanating from them". Bohr developed the concept of complementarity in quantum physics, which he employed as a philosophical concept and extended to biology. In his later years he turned his attention to nuclear physics and elementary particles. He also showed a keen interest in the developments of molecular biology.

Hans Bremer, b. 1927. In 1957 he was awarded a Ph.D. in zoology from the University of Göttingen, supervised by Karl Henke. The subject of study was blood vessel development. During 1957–59 he was scientific assistant at the Zoological Institute, Göttingen under Hansjochem

[82]Entry based on biographical information in: http://nobelprize.org/nobel_prizes/physics/laureates/1922/bohr-bio.html

Autrum; during 1959–62 he was an assistant at the Institute of Genetics, Cologne, with Carsten Bresch and Peter Starlinger, studying phage genetics and mRNA synthesis; in 1962–65 he was a research associate at the Virus Laboratory, University of California, Berkeley, with Gunther Stent, studying RNA polymerase function. From 1965–66 he was a scientific assistant, Institute of Genetics, Cologne, (Carsten Bresch), studying RNA polymerase function. He did not do a *Habilitation*, and then from 1966–69 he was assistant and associate professor, Southwest Center for Advanced Studies, Dallas, Texas. During 1969–95, Bremer was associate and full professor at the University of Texas, Dallas, with Richardson, and then during 1995–98, research scientist. In 1983 he was visiting professor in Geneva and in 1994, at the Academia Sinica, Taipei.

His post-doctoral research at Berkeley dealt with *in vitro* synthesis of RNA on T4 phage DNA templates, and properties of the transcriptions complex, direction and velocity of RNA chain growth, initiating nucleotides. Research at Dallas involved RNA synthesis, and chromosomal and plasmid DNA replication in *E. coli* bacteria (part of it summarised in a recent review: Control of rRNA synthesis in *E. coli*: a systems biology approach, *MMBR* 68:639–668).

Carsten Bresch, b. 1921. Bresch studied physics, and received his Dr.rer.nat. in 1955: He established a phage laboratory in Rio de Janeiro. In 1957 he received his *Habilitation* from Göttingen, and in 1957 was appointed professor (*Extraordinarius*) at Cologne. In 1962 he became head of the Biology Division and founder of the Graduate Research Center, University of Texas in Dallas. In 1964 he became professor at Freiburg. Bresch is the author of *Klassische und Molekulare Genetik* (1964), *Zwischenstufe Leben* (1977) and *Des Teufels neue Kleider* (1978).

Hermann Bujard, b. 1934. He received his Dr.rer.nat. from Göttingen in 1962; during 1964–65 he was a NIH post-doctoral fellow at the University of Wisconsin, Madison; 1966–69 he was assistant professor, Southwest Center for Advanced Studies, Dallas; 1970–82 professor of molecular genetics in Heidelberg; 1983–86 head of biological research, F. Hoffmann-La Roche & Co., Basel, Switzerland; 1986–99 founder and director of the ZMBH, University of Heidelberg; and since 1999 group leader, ZMBH Heidelberg. Awards: 1995 Beckurts-Preis; 1996 elected scientific member of the Max Planck Society; 1996 Prix Yvette Mayent for Cancer Research; 2000 honorary doctorate Würzburg. Research activities: control of gene activation at the transcriptional level in pro- and eukaryotes; malaria research.

Charles N. David, b. 1940. In 1962, David received his BA in biology from Harvard College, Cambridge, MA. He then spent one year at the Institute of Genetics, Cologne and in 1968 he received his Ph.D. from Caltech, Pasadena, CA, under Max Delbrück. His dissertation dealt with ferritin in the fungus *Phycomyces*. During 1968–73 he was a post-doctoral fellow at the MPI for Virus Research, Tübingen, advised by Alfred Gierer. At the MPI he worked on the developmental biology of Hydra, and discovered that Hydra could regenerate from reaggregated cells and demonstrated with this result that multicellular biological organisation could arise *de novo*. During 1973–81 he was assistant and associate professor, Department of Molecular Biology, Albert Einstein Medical School, New York, NY; during 1981–2005 he was a professor in the Department of Biology, University of Munich. Research activities: the role of cell proliferation and differentiation in relation to the morphology of Hydra; cloning of interstitial stem cells and the demonstration that they can give rise to a variety of differentiated cell types. He identified peptide signals stimulating nerve cell differentiation from stem cells and showed that nerve precursors migrate in tissue to sites of differentiation, thus giving rise to a spatial pattern of differentiated nerve cells. Work on growth control in Hydra led to the unexpected discovery that growth is regulated by cell death (apoptosis) and not cell proliferation.

Ute Deichmann studied biology at the universities of Bochum and Heidelberg. She taught biology from 1975–87. Dr.rer.nat.1991 (thesis on the history of biology) at the Institute of Genetics in Cologne (with Benno Müller-Hill); *Habilitation* in Cologne 2001 (thesis on the history of chemistry and biochemistry). Present position: Research professor at the Leo Baeck Institute London (since 2003). Other affiliations: working group leader ("History of the Biological and Chemical Sciences") at the Institute of Genetics in Cologne; research reader in intellectual history at the University of Sussex; regular visiting professor at Ben-Gurion University of the Negev, Israel; senior research fellow, Edelstein Center, Hebrew University, Jerusalem. Author of *Biologists under Hitler* (1992; 1996), *Flüchten, Mitmachen, Vergessen. Chemiker und Biochemiker im Nationalsozialismus* (2001), and numerous articles on the history of biology and chemistry. Co-editor of *Jews and Science in German Contexts. Case studies from the 19th and 20th Centuries* (2007). Recipient of the Ladislaus Laszt Award of Ben Gurion University (1995) and the Gmelin Beilstein Medal of the *Gesellschaft Deutscher Chemiker* (2005).

Walter Doerfler, b.1933. In 1959, he completed his medical dissertation at the University of München, Anatomisches Institut, under Titus von Lanz, "Über den funktionellen Einbau der vasa mesenterica". In 1961 received *Approbation,* and in 1996 became *Facharzt Humangenetik.* Post-doctoral positions: 1961–63 MPI für Biochemie, Munich, with Wolfram Zillig, on protein biosynthesis in a cell free system from *E. coli*; 1963–66, at the Department of Biochemistry, Stanford, with David Hogness, on separation of the strands of bacteriophage lambda DNA and generation of heteroduplex DNA molecules: reading strand for the lambda N gene. Assistant professor (1966–69); associate professor (1969–71); adjunct professor (1971–78), The Rockefeller University, New York; professor Institute of Genetics, Cologne (1972–2002); emeritus professor 2002; *Gastprofessor,* Erlangen since 2002. He has been visiting professor in Uppsala, Sweden, Stanford, Princeton, Vanderbilt, Kawasaki Medical School, Japan and Akademia Nauk, Russia. Major scientific results: Integration of adenovirus type 12 (Ad12) DNA in hamster cells and tumours (1968 to present); abortive infection of hamster cells by Ad12 (1969 to present); *de novo* methylation of integrated foreign DNA in mammalian cells (1978); inverse correlations between promoter methylation and gene expression (1979, 1980, 1983); promoter methylation inactivates eukaryotic genes (1982); transcription profiles of baculoviruses (1984); spreading of *de novo* methylation (1989); DNA methylation patterns in different parts of the human genome (starting 1990); fate of food ingested foreign DNA in the gastrointestinal tract (1994); DNA methylation in the promoter region of the human FMR1 gene (2000, 2006).

Horst W. A. Feldmann, b.1932 in Stettin. He received his Dr.rer.nat. at Cologne under Leonhard Birkofer in organic chemistry; during 1962–67 he collaborated with Hans-Georg Zachau at the Institute of Genetics in Cologne on characterisation of the aminoacyl-tRNA bond (2′-3′ problem); nucleotide sequences of two serine-specific transfer ribonucleic acids from brewer's yeast; and identification of three "minor" nucleotides in these tRNAs. In 1968 he received the *Habilitation* (thesis on multiplicity of tRNAs in brewer's yeast), University of Munich, *venia legendi* in biochemistry. During 1974–97 he was professor at the Institute of Physiological Chemistry (Medical Faculty), Munich. Present position: Professor emeritus, molecular biology at Adolf Butenandt Institut, Munich. Research: multiplicity of tRNAs in yeast (sequencing of minor serine-specific, methionine-specific, glutamic acid-specific tRNAs); characterisation of yeast tRNA genes, their abundance, precursors and expression; tRNAs in mitochondria; sequencing of rat liver mitochondrial DNA; characterisation of transposable elements in yeast (Ty-elements) and their affiliation with tRNA genes; and genome organisation in yeast.

Rainer Hertel, b. 1937. He studied biology, from 1956, at the University of Munich, and undertook doctoral research from 1960–62 on auxin transport in the laboratory of A. Carl Leopold, Purdue University, Lafayette. In 1963 he received his Dr.rer.nat. from Munich; during 1963–66 he was a DFG-post-doctoral fellow and worked on phage recombination with Carsten Bresch at the Institute of Genetics in Cologne. During 1966–69 he was assistant/associate professor at the MSU/AEC Plant Research Laboratory, East Lansing, Michigan, USA, where he did research on auxin and plant gravitropism. In 1969 he became professor at the Institute Biologie III, Freiburg. His research focussed on auxin binding, gravi- and phototropism in plants and on theoretical biology. He taught genetics, philosophy of science and history of biology. For many years, he served as director of the institute or as chairman of the faculty. He retired in 2000, and continues work on some left-over auxin and phototropism questions, on aspects of human colour vision (how can red/green-blind men help to understand the evolution of colour vision?) and on some philosophical problems of biologists.

Jonathan C. Howard, b. 1943. During 1961–64 he studied for his BA in zoology at Oxford University. During 1964–65 he was recipient of the Royal Society Leverhulme Award, working at the Genetics and Biometry Laboratory (laboratory of J. B. S. Haldane and S. D. Jayakar), Government of Orissa, Bhubaneswar, India; during 1965–69 he undertook doctoral research, supervised by J. L. Gowans, MRC Cellular Immunology Unit, Sir William Dunn School of Pathology, Oxford, where, in 1969, he received the DPhil (Medicine) on "Cellular Aspects of Antibody Formation"; during 1969–73 he was a member of MRC scientific staff; during 1970–73 he was Weir Junior Research Fellow in Science, University College, Oxford; 1971–72, at the University of Pennsylvania School of Medicine, Department of Pathology; 1973–78, adjunct, assistant, and associate professor, School of Medicine, Department of Pathology, University of Pennsylvania; 1974–94, member, scientific staff, Biotechnology and Biological Sciences Research Council, Immunology Department, The Babraham Institute, Cambridge, UK; 1985–94, head, Immunology Department, Babraham Institute; 1994 to present, professor, Institute of Genetics in Cologne. Since 1995 he has been a Fellow of the Royal Society. Research Activities: Immunogenetics. He has worked on the mechanism of antigen presentation and was one of the discoverers of the Tranporter associated with antigen processing (TAP). He now works on interferon-induced proteins that are responsible for the destruction of intracellular bacterial and protozoal pathogens. He has also maintained an interest in evolutionary mechanisms since his stay in India and has written a book on Darwin's evolutionary theories (*Darwin. A Very Short Introduction,* 1982 and 2001).

Lothar Jaenicke, b. 1923 in Berlin. *Abitur* 1941 (Frankfurt). Prevented from regular studies by Nazi laws, he was auditor in biology and chemistry from 1941 to 1943 (U. Marburg) and *Laborant* (assistant chemist) from 1943 to 1945 at Schering AG, Berlin. He studied medicine and chemistry in Marburg, where he wrote his diploma thesis in chemistry (1948) and dissertation (Dphil, oxoniumsalt complexes of tetrahydrofurane) with H. Meerwein. 1949 he became research assistant (chemistry) in Tübingen and from 1949 to 1957 in Marburg (physiological chemistry, biochemistry (self taught): function of tetrahydrofurane in C_1-metabolism). In 1953 he received his *Habilitation* in general chemistry. From 1953–56 he was research fellow at the Department of Biochemistry, Western Reserve University, Cleveland, Ohio (polishing up biochemistry). 1957–62 he was *Dozent* and apl. Professor at the Biochemistry Institute, University of Munich. During 1962–63 he was a.o. professor of physiological chemistry at the medical faculty in Cologne; 1963–88 professor of biochemistry, science faculty, where he was founder of the Institute of Biochemistry. 1988 he became emeritus professor and was visiting research scientist and professor at various institutions: Ain Shams University, Cairo; AUB, Beirut; IIS, Bangalore; AIMS, New Delhi; Department of Botany, UTEX, Austin. From 1986–87 he was a fellow of the *Wissenschaftskolleg zu Berlin*.

Awards: Paul Ehrlich/Ludwig Darmstaedter-Preis; Warburg, Lorenz Oken, Richard Kuhn Medals. Research activities: Cofactor functions and enzyme mechanisms in tetrahydrofolate and cobalamine-catalysed metabolic reactions; alternative metabolic ways leading to one-carbon fragments; methyl group transfers to and from methionine; fatty acid derived pheromones and signal compounds in brown and green algae; differentiation and cell wall lysis in Volvocales algae; natural products from isoprenoids and their role in signal chains as attractants and as developmental triggers. Founder of the journal, *BIOspektrum*, and columnist of the series "Portraits for Memory in Biosciences".

Hubert Kneser. He studied chemistry, including mathematical and biological topics in Tübingen and Munich, beginning in 1949. His diploma and doctoral theses in physical chemistry were on questions of the kinetics of reactions occurring at the boundary surface at solid catalysts. After two post-doctoral years at the MPI of Virus Research with Hans Friedrich-Freksa and Fritz Kaudewitz, where he conducted preliminary experiments concerning bacterial survival after P-32 incorporation and enzyme synthesis under different culture conditions, he found, during the first phage course in the new Cologne Genetics Institute, a place in Peter Starlinger´s lab. Here, by UV irradiation of bacteria of diverse genetic constitution in media stimulating as well as inhibiting DNA repair, he could clarify some of the intricate interactions of several repair mechanisms regarding bacterial survival, as well as induction of prophage. From 1967 to 1969 he stayed with George Streisinger at the University of Oregon in Eugene, trying to find correlations between mutability and surrounding nucleotide sequences in defined places of the T4 lysozyme gene, whose sequence was already established at that time. Incidentally Kneser stumbled over a suppressor-tRNA gene within T4. Back in Cologne he turned to measuring stability and kinetics of formation and decay of RNA-polymerase complexes with well-known promoters, but gradually shifted to teaching and administrative tasks.

Joseph W. Lengeler, b.1937. In 1966 he did his dissertation with Peter Starlinger, Institute of Genetics, Cologne, "Analysis of the glucose effect on the synthesis of the galactose enzymes in *Escherichia coli*". Post-doctoral positions: 1966–69 with Peter Starlinger, Cologne, work on galactose transport systems in *E. coli* and their role in chemotaxis; 1969–72 with E. C. C. Lin, Harvard Medical School, Boston, work on genetics and biochemistry of hexitol-specific PEP-dependent phospho-transferase systems (PTS) from enteric bacteria; their glycerol metabolism; 1972–73 with Starlinger, Cologne; 1973–75 with W. Tanner, Institut für Botanik, U. Regensburg; 1975–79 with R. Schmitt, Institut für Genetik, U. Regensburg. In 1976, *Habilitation* in genetics, Fachbereich Biologie, Regensburg; 1980 professor (C2) of genetics, Regensburg; 1984 professor (C4) of genetics, Fachbereich Biologie/Chemie. Genetics and biochemistry of various PTSs from enteric bacteria; their role in transport, catabolite repression, inducer exclusion, and in chemotaxis; elucidation of the corresponding catabolic pathways; studies in experimental evolution of complex pathways and the role of autonomous genetic elements, e.g., conjugative transposons and genomic islands; since 1990, work in systems biology with groups in Stuttgart; since 2002, *Gastprofessor* at the MPI für Dynamik komplexer technischer Systeme in Magdeburg. Mathematical and computer modelling of complex biological systems. Author of *Biology of the Prokaryotes* (1999).

Maria Leptin, b. 1954 in Hamburg. In 1979, degree in mathematics and biology, Heidelberg; 1979–83 dissertation with Fritz Melchers at the Basel Institute for Immunology, degree 1983, Heidelberg; 1984–87 post-doctoral fellow with Michael Wilcox at the MRC Laboratory for Molecular Biology, Cambridge, UK; 1988 staff scientist MRC LMB, Cambridge; Jan-Mar 1989 guest scientist with P. O'Farrell, University of California, San Francisco; 1989–94 research group leader at the MPI for Developmental Biology, Tübingen; 1994 to present, professor Institute of Genetics, Cologne; editorial board *Developmental Cell*; co-editor of *Mechanisms of Development*; Chair of EMBO membership council. Research activities: Developmental biology in *Drosophila* and zebrafish.

Georg Friedrich (Fritz) Melchers, b. 27 April 1936 Berlin. Degrees: l96l Dipl. chem., Cologne; 1964 Dr.rer.nat., Cologne; 1971 Dr.rer.nat. Habil., *venia legendi,* Freiburg, Faculty of Biology. Positions held: l965–67 research associate, Salk Institute for Biological Studies, La Jolla, California (Fulbright Travel Scholar); l968–69 post-doctoral fellow, 1970 senior research assistant MPI for Molecular Genetics, Berlin. Visiting scientist, Weizmann Institute, Rehovot, Israel (1970), department of genetics, Stanford Medical School (1970–71). 1970–80 permanent member, Basel Institute for Immunology, Basel; 1980–2001 director of the Institute; 2001

Institute was closed down by Roche; 1981 to present, extraordinary professor of immunology, Basel; 2003 to present, senior research group leader, MPI for Infection Biology, Berlin.

At the Basel Institute for Immunology Melchers selected, supported, guided, critically reviewed and integrated over 300 scientists with their research projects in most major fields of immunology. In 2002, he was co-founder of 4-Antibody, a company that intends to produce and improve human antibodies. In 2003, he became head of a Senior Research Group at the MPI for Infection Biology with which he intends to continue some of the research on stem cells and B cell development. From 1973–79 he gave lectures and practical courses in immunology at the Faculty of Biology in Heidelberg, and from 1980 until present in Basel.

Among the many advisory and review activities of Melchers are the following: In 1985 the Government of Berlin asked him to help in the planning of a biomedical research centre for rheumatoid arthritis. This has led to the foundation and successful establishment of the *Deutsche Rheumaforschungszentrum* (DRFZ), housed at the Charité, together with the MPI for Infection Biology, and engaged in research on autoimmune diseases in close collaboration with the Chair of Rheumatology. Melchers is a member of the *Stiftungsrat* of the DRFZ.

In 1990 the German Science Council (*Wissenschaftsrat*) asked Melchers to participate in the review of 17 biomedically oriented institutions of the Academy of the former German Democratic Republic, which resulted in the establishment of a new foundation for research in molecular medicine in Berlin-Buch, the Max Delbrück Center for Molecular Medicine.

As permanent scientific member and director of the Basel Institute for Immunology (which was an academically free research institute, wholly supported by F. Hoffmann-La Roche), Melchers restricted his consultancy for commercial companies to giving Roche, especially in the areas of gene expression and protein production, monoclonal antibodies in diagnosis and therapy, cytokines and cytokine receptors (Il-1, TNF-a, etc.) in inflammation and autoimmune diseases, and vaccine developments, especially against influenza and malaria. After his retirement he began consulting for law firms and biotech companies.

Georg Friedrich Michaelis, Dr.rer.nat., Universitätsprofessor i.R., born 1938. He studied chemistry, from 1958, in Cologne; 1964: Dipl.-Chem. 1964–68 he undertook doctoral work on the expression of the galactose operon in *E. coli.* at the Institute of Genetics in Cologne with Peter Starlinger. Post-doctorate positions: 1968–69 Institute of Genetics, lab of Peter Starlinger, work on IS elements in the gal operon; 1969–71 Department of Biochemistry, University of California at Davis, lab of R. S. Criddle, work on mitochondrial transcription and yeast petite mutants without mitochondrial (mt) DNA (rho zero mutants); 1971–73 Centre de Génétique Moléculaire du CNRS, Gif-sur-Yvette, France, lab of Piotr Slonimski, work on recombination of mtDNA; 1973–77 Institute for Physiological Chemistry, Würzburg, lab of E. Wintersberger, work on mitochondrial antibiotic resistance mutations in yeast. 1977 *Habilitation* for genetics, Würzburg, work on mitochondrial antimycin resistance in yeast. 1977–84 *Privatdozent,* later apl. professor of molecular biology, Bielefeld; 1984–2003 professor, Abteilung Botanische Cytologie, Düsseldorf.

Major scientific results: cytoplasmic male sterility in *Beta maritima,* sequence of mitochondrial DNA from *Chlamydomonas reinhardtii,* recombination of mtDNA from yeast, mitochondrial transcription machinery, mitochondrial protein transporter Oxa 1, mitochondrial signalpeptidases (the inner membrane peptidase Imp1, the rhomboid peptidase Pcp1).

Bernhard Mühlschlegel, b. 13 September 1925 in Berlin. 1953 Dr.rer.nat. Humboldt University of Berlin, theoretical physics; scientific assistant in Berlin, Heidelberg and Munich; 1960 *Habilitation* University of Munich; 1960–62 research assistant professor, University of Illinois; 1962–90 professor of theoretical physics, Cologne; 1990 emeritus professor. Numerous research activities in condensed matter physics, also during sabbaticals at FU Berlin, and the Universities of Pennsylvania, California, Stanford, and Tel Aviv; 1991–98 Chief Editor of *Annalen der Physik*.

Benno Müller-Hill, b.1933 in Freiburg. I. B., studied chemistry in Freiburg and Munich; Ph.D. under Kurt Wallenfels with work on *Struktur und Wirkungsmechanismen der Alkoholdehydrogenase aus Bäckerhefe und der ß-Galactosidase aus E. coli ML309.* Post-doc 1963–64 with Howard Rickenberg in Bloomington, Indiana and 1965–68 with Walter Gilbert at Harvard University. Together with Walter Gilbert he isolated 1966 Lac repressor (LacR), the first transcription-factor. He also isolated a mutant which produced one hundred times more LacR than the wild type. In 1968 he became professor at the Institute of Genetics in Cologne. There he and his collaborators analysed structure and function of LacR. In particular they solved the function of the auxiliary operators 02 and 03: they are present to increase local concentrations of LacR around 01. He also worked on malaria and Alzheimer's disease. He published three books: *Die Philosophen und das Lebendige, Murderous Science,* and *The lac Operon.* He received an honorary doctoral degree from the Technion in Haifa. After his retirement he is still active in theory and history of molecular biology.

Karl Wolfgang Mundry, b. 16 March 1927 in Hildesheim, studied botany at the university in Göttingen from 1946–49. In 1950, he moved to Tübingen to work on plant viruses with Georg Melchers at the MPI für Biologie. In 1954, he received his Ph.D. at the University of Tübingen with a thesis on the *in vitro* and *in vivo* mutability of TMV. In the fall of the same year, he became head of the virology department of the "Institut für Landwirtschaftliche Technologie und Zuckerindustrie an der T. H. Braunschweig" to work with sugar beet viruses (beet yellows and beet mosaic). In 1957, he accepted an assistant position with G. Melchers at the MPI für Biologie in Tübingen. From 1959–61, he spent two years at the Biology Division of the California Institute of Technology in Pasadena, California with Robert Sinsheimer. Returning to the MPI für Biologie, he became a staff member in 1968. In 1972, he was appointed Professor of Botany at the University in Stuttgart and, in 1993, he became Professor Emeritus.

Peter Overath, b. 1935, studied chemistry at the Universität München; 1961 dissertation with F. Lynen on the role of vitamin B_{12}-coenzyme in the rearrangement of methylmalonyl-CoA to succinyl-CoA; post-doctoral stay at Davis, California, with P. K. Stumpf; 1966–73 research assistant and group leader at the Institute of Genetics, Cologne (*Habilitation* 1969) on the genetics of fatty acid degradation in *E. coli* and the structure-function relationship of phospholipids in biological membranes; 1973–2003 director at the MPI for Biology in Tübingen; 1974 honorary professor in Cologne; since 2003 emeritus professor, Department of Immunology, Tübingen. Research activities: phospholipids and characterisation of the proton-galactoside co-transporter lactose permease of *E. coli*; 1980 sabbatical stay in London with G. A. M. Cross at the Wellcome Laboratories; analysis of the surface organisation and intracellular membrane transport in *Trypanosoma* and *Leishmania,* influence of the MHC on behaviour of mice.

Klaus Rajewsky, b. 1936. Studied medicine in Frankfurt and Munich, research stay at the Institut Pasteur in Paris; 1962 medical dissertation, Frankfurt; 1964–70 research assistant at the Institute of Genetics in Cologne; 1970–2001 professor; 2001 emeritus professor; 2001 professor of pathology, Harvard Medical School, Center for Blood Research, Cambridge, Massachusetts. 1967 founding member of the German Immunological Society; 1995–2001 head of the EMBL Monterondo, Italy; 1998 co-founder of Artemis Pharmaceuticals. Awards: 1977 Avery-Landsteiner Award of the *Deutsche Gesellschaft für Immunologie*; 1994 Behring Kitasato Prize; 1996 Robert-Koch-Preis with Fritz Melchers; 1997 Körber-Preis; 2001 Deutscher Krebshilfepreis with Martin-Leo Hansmann and Ralf Küppers; 2004 Dr. h. c. Goethe-U. Frankfurt; 2005 Brupbacher Prize for Cancer Research with Mariano Barbacid. Research activities: immunogenetics in mice and humans; mutant mice as model organisms for genetic research; B-cell differentiation.

Heinz Saedler, b. 3 June 1941 in Bad Godesberg. 1960–67 study of chemistry, biochemistry and genetics in Bonn, Munich and Cologne; 1967 Dr.rer.nat. in genetics; 1967–69 assistant in Cologne; 1969–70 post-doc at the Department of Biological Sciences, Stanford; 1974 *Habilitation* in genetics, Cologne; 1975–80 associate professor Freiburg; since 1980 director of molecular plant genetics at the MPI for Plant Breeding Research; since 1981 honorary professor University of Cologne. Awards: 1985 Otto Bayer Preis; 2000 Wilhelm Exner Medaille. Research activities: transposable elements in plants, flower development and evolution, mechanisms of evolution of plants. 1979–99 editor of *Molecular and General Genetics.*

Peter Starlinger, b. 1931. 1954 Dr. med. in Tübingen; doctoral work with Hans Friedrich-Freksa at the MPI for Biochemistry; later MPI for Virus Research; 1960 *Habilitation* (genetics of micro-organisms) in Cologne; 1965 professor for genetics and radiation biology at the Institute of Genetics in Cologne; 1996 emeritus professor. Awards: 1979 Robert-Koch-Preis, 1985 Otto-Warburg-Medaille, 1986 Feldberg-Preis, Ernst-Hellmut-Vits-Preis, 1990 Theodor-Boveri-Preis. Research in molecular biology, especially on transposable DNA elements in bacteria and plants.

Gunther S. Stent b. 1924 in Berlin. He fled from Nazi Germany in 1938 and completed his secondary education in Chicago. Stent studied physical chemistry at the University of Illinois (BS 1945; Ph.D. 1948), and then went to the California Institute of Technology as a post-doctoral fellow of the National Research Council, to join Max Delbrück's "Phage Group", the fountainhead of the discipline that, a few years later, came to be called Molecular Biology. Stent has been on the faculty of the University of California, Berkeley since 1952, as professor of molecular biology since 1952, as Founding Chairman of the Department of Molecular and Cell Biology from 1980–92, and as professor emeritus of neurobiology since 1994. Besides contributing to the scientific literature and being the author of several textbooks, Stent has also published on the history and philosophy of science, as well as an autobiographical memoir, *Nazis, Women and Molecular Biology.* He is a member of the

United States National Academy of Sciences, the American Academy of Arts and Sciences, and the American Philosophical Society, as well as an external member of the German Max Planck Society.

Bruno J. Strasser, b. 1972, is a historian of science and medicine, currently assistant professor at Yale University. He holds a Ph.D. from the Universities of Paris 7 and Geneva. Bruno J. Strasser. He is the author of a book on the history of molecular biology in Switzerland (*La Fabrique d'une Nouvelle Science, La Biologie Moléculaire à l'Age Atomique, 1945–1964* (2006)), and of numerous articles in *Science, Nature,* and various historical journals. He has also edited two collected volumes, one on *Molecular Biology in Postwar Europe* (together with Soraya de Chadarevian, in *Studies in the History and Philosophy of Biological and Biomedical Sciences*, 2002, 33C) and one on *Science, Industry and the State in the 20th Century* (together with Michael Bürgi, in *Schweizerische Zeitschrift für Geschichte,* 2005, 55, 1). He has been a visiting fellow at the University of Princeton, at the MPI für Wissenschaftsgeschichte in Berlin and at the Ecole Normale Supérieure in Paris, and a senior lecturer and researcher in Geneva and Lausanne. Strasser is currently working on the history of biomedical collections in the 20th century, from the American Type Culture Collection to GenBank, on drug innovation in the pharmaceutical industry, on science and foreign policy during the Cold War, and on the legal history of paternity.

Joseph Straub (1911–1987) studied natural sciences at the University of Freiburg. 1930–39 Ph.D. and *Habilitation* with Friedrich Oehlkers, Freiburg. Until 1945 he worked at the Kaiser Wilhelm Institute for Biology in Berlin with Fritz von Wettstein, from 1945–49 at the same Institute, (from 1948 MPI), in Tübingen. In 1949 Straub became head of the Institute of Botany in Cologne and, in 1961, director at the MPI for Plant Breeding Research in Cologne-Vogelsang. Straub initiated and co-founded the Institute of Genetics in Cologne.

Thomas A. Trautner, b. 3 April 1932, Göttingen, Germany. Education: studied biology, microbiology, biochemistry and genetics in Münster and Göttingen and at the University of Illinois. Degrees: 1957 Dr.rer.nat., Göttingen; 1963 *Habilitation*, Cologne; 1997 Dr. med.h.c., Humboldt University, Berlin. Fellowships: 1953–54 Fulbright Commission, Exchange Fellowship for Undergraduate Studies, University of Illinois; 1957–59 fellow (DFG) Institute of Genetics, Cologne; 1959–61 post-doctoral fellow (DFG) Department of Biochemistry (with A. Kornberg), Medical School, Stanford. Positions held: 1961–63 research assistant, Institute of Genetics, Cologne; 1963–65 assistant professor, Virus Laboratory and Department of Molecular Biology, University of California, Berkeley; 1965–2000 director at the MPI for Molecular Genetics, Berlin; 1967–2002 honorary professor, Freie Universität Berlin; 1990–96 vice-president (biology/medicine) of the Max Planck Gesellschaft; member and/or chairman of some major scientific boards.

Simone Wenkel, b. 1977. Studied biology in Mainz, Würzburg and Cologne and finished her diploma thesis at the MPI for Neurological Research in Cologne in 2003. She is working on her doctoral thesis on the history of molecular biology in the "History of the Chemical and Biological Sciences" group of Ute Deichmann at the Institute of Genetics in Cologne and is currently a visiting graduate student at the History and Philosophy of Science and Technology Program at Stanford University.

Hans-Georg Zachau, b. 16 May 1930 in Berlin. 1948–50 study of preclinical medicine in Frankfurt, *Physikum*; 1948–53 study of chemistry in Frankfurt, diploma exam; 1952–55 research for diploma and doctoral theses on the sex attractant of the silk moth at the MPI für Biochemie in Tübingen under Adolf Butenandt; 1955 Dr.rer.nat. in chemistry; 1955–58 post-doctoral work at the MPI für Biochemie (synthetic chemical work), MIT, Cambridge, Massachusetts (work on the structure of the cyclic peptide antibiotic Etamycin (J. C. Sheehan)), and the Rockefeller University, New York (work on protein biosynthesis and tRNA (F. Lipmann)). 1958–61 independent research work at the MPI für Biochemie, Munich, 1961–66 group leader at the Institute of Genetics in Cologne. 1962 *Habilitation* in physiological chemistry; 1965/66 offers of professorships for physiological chemistry in Berlin, genetics in Cologne, and physiological chemistry in Munich; since 1967 professor of physiological chemistry at the University of Munich and co-director of the Institute for Physiological Chemistry (since 1998 Chair for Molecular Biology). 1999 emeritus professor; 1981 member of the *Orden Pour le mérite*; 1992 chancellor. Research activities: 1957–81 tRNA structure and interactions, protein biosynthesis; 1970–84 repetitive DNA, chromatin; since 1977 immunoglobulin genes.

Abbreviations

AcNPV	*Autographa californica* (nuclear polyhedrosis virus)
Ad12	adenovirus 12
ATP	adenosine triphosphate
BASF	Badische Anilin- und Sodafabrik
BSE	bovine spongiform encephalopathy
C4/C3	salary grades of German professors
cAMP	cyclo adenosine monophosphate
CEO	chief executive officer
CFU	cellular functional unit
CMMC	Center for Molecular Medicine Cologne
CRP	cAMP receptor protein
CsCl	caesium chloride
dept.	department
DFG	*Deutsche Forschungsgemeinschaft*
DNA	deoxyribonucleic acid
E. coli	Escherichia coli (a non-pathogenic colon bacterium)
EMBL	European Molecular Biology Laboratory
EMBO	European Molecular Biology Organisation
Erüpa	*Entrümpelungsparty* (clearing out party)
EU	European Union
gal	galactose
hets	heterozygotes
HMS	Harvard Medical School
IBch/UKln	Institute for Biochemistry at the University of Cologne
IIA^{Crr}	component of the PTS
IS	insertion element

K-12	*E. coli* strain
KWI	Kaiser Wilhelm Institute
lac	lactose
LMB	Laboratory for Molecular Biology, Cambridge
M13	bacterial vector
MHC	Major histocompatibility complex
MIT	Massachusetts Institute of Technology
MNU	*Landesstelle für den Mathematisch-Naturwissenschaftlichen* Unterricht (German teaching administration unit)
MPG	Max Planck Society
MPI	Max Planck Institute
MPIZ	Max Planck Institute for Plant Breeding Research
MRC	Medical Research Council, United Kingdom
mRNA	messenger RNA
NIH	National Institute of Health
NRW	North Rhine-Westphalia
P32	radioactive phosphorus
PCR	polymerase chain reaction
PEP	phosphoenol pyruvate
Ph.D.	Philosophical Doctor
PISA	Programme for International Student Assessment
PNAS	Proceedings of the National Academy of Sciences, USA
PTS	phosphotransferase system
RNA	ribonucleic acid
SFB	*Sonderforschungsbereich* (Collaborative Research Centre) funded by the DFG
T1 or T4	bacteriophage strains
TMV	tobacco mosaic virus
tRNA	transfer RNA
trp	tryptophane
UTD	University of Texas, Dallas
UV	ultraviolet
ZMBH	Center for Molecular Biology, Heidelberg
λ	λ bacteriophage
ΦX174	bacteriophage

Figure Sources

The figures in this book were kindly provided by the following sources:

Fig. 1–2	Archives of the Max Planck Society
Fig. 3	Institute of Genetics, Cologne
Fig. 4	Kölner Stadtanzeiger
Fig. 5–6	Bildarchiv Volkswagenstiftung
Fig. 7–10	Charles N. David
Fig. 11–13	Horst Feldmann
Fig. 14–23	Fritz Melchers
Fig. 24	Hans Bremer
Fig. 25	Walter Doerfler
Fig. 26	Rainer Hertel
Fig. 27–28	Institute of Genetics, Cologne
Fig. 29–34	Niko Haase
Fig. 35–42	Brigitte von Wilcken-Bergmann

Index of Names